SpringerBriefs in Applied Sciences and Technology

SpringerBriefs present concise summaries of cutting-edge research and practical applications across a wide spectrum of fields. Featuring compact volumes of 50 to 125 pages, the series covers a range of content from professional to academic.

Typical publications can be:

- A timely report of state-of-the art methods
- An introduction to or a manual for the application of mathematical or computer techniques
- A bridge between new research results, as published in journal articles
- A snapshot of a hot or emerging topic
- An in-depth case study
- A presentation of core concepts that students must understand in order to make independent contributions

SpringerBriefs are characterized by fast, global electronic dissemination, standard publishing contracts, standardized manuscript preparation and formatting guidelines, and expedited production schedules.

On the one hand, **SpringerBriefs in Applied Sciences and Technology** are devoted to the publication of fundamentals and applications within the different classical engineering disciplines as well as in interdisciplinary fields that recently emerged between these areas. On the other hand, as the boundary separating fundamental research and applied technology is more and more dissolving, this series is particularly open to trans-disciplinary topics between fundamental science and engineering.

Indexed by EI-Compendex, SCOPUS and Springerlink.

Endre Magoss · György Sitkei · Zoltán Kocsis

Dust Extraction and Handling in the Wood Industry

Springer

Endre Magoss
Department of Wood Engineering
University of Sopron
Sopron, Hungary

György Sitkei
Budapest, Hungary

Zoltán Kocsis
Department of Wood Engineering
University of Sopron
Sopron, Hungary

ISSN 2191-530X ISSN 2191-5318 (electronic)
SpringerBriefs in Applied Sciences and Technology
ISBN 978-3-031-08914-5 ISBN 978-3-031-08915-2 (eBook)
https://doi.org/10.1007/978-3-031-08915-2

This Springer imprint is published by the registered company Springer Nature Switzerland AG
The registered company address is: Gewerbestrasse 11, 6330 Cham, Switzerland

Who does not know the past, that cannot value the present.

W. Goethe (1749–1832)

Preface

Extraction and handling of dust and chips is an indispensable part of any production line in the wood industry. There are strict national and international regulations and prescriptions concerning air quality in workrooms, safety measures for avoiding fire and dust explosion, which are highly dependent on the efficiency of dust extraction from woodworking machines and dust separation devices, physical-mechanical properties of dust particles, and their behavior in air stream and bulk.

This book is an attempt to summarize our present knowledge on this subject, and also several new derivations, experimental results, and new suggestions are introduced.

The main disadvantageous properties of wood dust and chips are their irregular shape, hairy surface with voids, and their more or less plastic behavior during static and impact loading. In bulk, they are inclined to settling due to their own weight (creeping) which makes difficult the outflow through an orifice. In order to overcome these difficulties, it is necessary to know their important physical and mechanical properties and to use them correctly to calculate and optimize technological processes. Some new experimental results and theoretical derivations stimulate us to revise till now used calculation methods such as particle separation in cyclones. Furthermore, a quite new theory and calculation method is elaborated describing friction losses in pneumatic transportation of wood dust and chips.

Numerous improvements are included to the topic such as

- new measurement results for particle size distribution as a function of wood species and machining,
- terminal velocity of small wood particles, comparison of measured and calculated values,
- new measurements for compaction behavior of dust and chips, and calculation of settling in storage bins,
- aerodynamics of small particles and their throwing distance as a function of terminal velocity,
- derivation of a new general equation for slip calculation in the case of impact losses.

- energy consumption and transportation efficiency illustrated with worked examples,
- a detailed analysis of particle separation efficiency including the new particle size distribution after separation,
- new calculation method for transport ventilators as a function of dust concentration,
- the interaction of ventilation and air recirculation is analyzed in detail and illustrated with worked examples.

Another feature of this book is also the mention and use of the valuable theoretical and experimental results of the past which serve, as a fundamental starting point, for further developments.

It is our belief that this material based on scientific approaches will serve a real need for a long time for all those working and teaching in the field covered by this work. The authors are grateful to the staff of Springer Verlag for their excellent co-operation.

Sopron, Hungary
February 2022

Endre Magoss
György Sitkei
Zoltán Kocsis

Introduction

The different woodworking operations in the wood industry are associated with dust and chip formation. The sizing, shaping, and finishing processes produce considerable quantity of dust and chips which may amount to 20–30% of the initial raw material, but, in some cases such as profiled window frames, it may be up to 50%. The produced dust and chip must be collected and transported in a storage bin for further utilization. The finer dust fraction is airborne and may pollute the air in the working space causing health risks. In order to avoid the spread of fine particles into the working place, the best and only method is the use of suction devices (hoods) just at the origin. The suction hood encloses the tool and the produced particles will be captured and carried away by the air stream. Due to the depression inside the hood, the particles do not get out into the surrounding air.

The handling of collected dust and chips is not a simple task. The wooden particles have a rough surface, irregular shape, and a given size distribution. The particles may stick together (coagulation) and in bulk, they are not free-flowing. Therefore, an appropriate handling requires the knowledge of physical-mechanical properties of particles and their mechanical behavior in bulk.

An important task is the separation of dust particles from the conveying air using different filter equipment. Using ventilation system with air recirculation, not only the overall filtration efficiency is needed but also the particle size distribution of filtered air recirculated into the workroom. Due to the strict regulation for the allowable dust concentration in workroom, detailed input parameters and careful design are required to comply with regulations in an economic manner.

Bulk granular materials have distinct behavior in many aspects. At the same time, early experiments (mid 1920s) in civil engineering have shown that "it is rather remarkable that the theory developed by Boussinesq for elastic solids fit so closely when applied to granular material" (Griffith 1931, 1934). That means that the basic behavior of granular materials follows well the main stress and deformation regularities of the homogeneous and elastic infinite or finite half-space.

While the theory of pneumatic conveying for impact losses (crop products) is long time well established (Pápai 1954), the same for friction losses (wood dust and chips) is much less elaborate. Therefore, attempts are made to analyze the friction losses and derive new relationships based on scientific approaches.

The first detailed experiments on pneumatic conveying have been carried out almost 100 years ago and the proposed simplified calculation method is widely used also today (Gasterstädt 1924). This calculation method uses a "constant" k_0, which is actually depending on various system variables. Using now the derived relationships, the constant k_0 can be expressed with the influencing variables and, with this completion, the Gasterstädt's equation can now be regarded as a generally valid calculation method.

New calculation methods are worked out for determining the filter separation processes including the particle size distribution in the air passed the filter.

In the following chapters a detailed discussion is given concerning the most important material properties, the mechanical behavior in bulk, the aerodynamics of dust extraction, the pneumatic conveying, filter separation processes, air recirculation, auxiliary equipments, and finally health and safety problems.

References

L. Gasterstädt, Die experimentelle Untersuchung des pneumatishen Förderganges. VDI Forschungsarbeiten, No. 265 (1924)

J. Griffith, Physical properties of earths. Iowa Eng. Exp. Stat. Bull. **101**, Ames (1931)

J. Griffith, Dynamics of earth and other macroscopic matter. Iowa Eng. Exp. Stat. Bull. **117**, Ames (1934)

L. Pápai, Pneumatikus gabonaszállítás (Pneumatic grain conveying) Magyar Tudományos Akadémia VI. Oszt. Közlemény, 12. pp. 319–363 (1954)

Contents

Chapter 1
Physical Properties of Dust and Chips

Physical properties describe the shape, size distribution of dust particles, their friction properties, terminal velocity, shear stress–normal stress relationships, and specific surface. Most of these properties are unique for wood dusts due to their originating mechanism (brittle fracture) and the porous nature of solid woods. We treat here only those properties which influence the basic processes covered by this work. Most of these properties can only be determined experimentally.

1.1 Origin of Dust and Chips

The dust and chips are produced by the different woodworking operations such as sawing, planing, milling, molding, sanding, etc. Depending on the operational characteristics such as cutting speed, tooth bite, depth of cut, and, on the other hand, on the mechanical properties of wood species (density, moisture content, resin content) quite different dust and chips may be obtained, concerning their size distribution and physical properties. The expected quantity by volume of dust and chips produced in the unit time is determined by the width and depth of cut and the feed speed. The quantity by mass (kg/s) is obtained by multiplying the volume with the wood density:

$$m_m = \frac{b \cdot e \cdot H}{60} \cdot \rho_m \quad \text{kg/s} \tag{1.1}$$

where

- b is the width of cut,
- H is the depth of cut,
- e is the feed speed, m/min,
- ρ_m is the wood density.

E. Magoss et al., *Dust Extraction and Handling in the Wood Industry*,
SpringerBriefs in Applied Sciences and Technology,
https://doi.org/10.1007/978-3-031-08915-2_1

In order to suck away all particles in the hood, the quantity of air (m^3/s) is generally prescribed by the machine manufacturer based on experiments. These quantities may vary in a broad range depending on the hood construction. In general, quantities from 600 to 1500 m^3/h (0.17–0.42 m^3/s) may be encountered. The chip production rate rarely exceeds the value of 0.1 kg/s and, therefore, the transport concentration in the tube is generally less than 1.0 kg/kg (see in Sect. 4.2).

The loose bulk density of dust and chips depends on:

- the solid wood density from which it is produced,
- the particle size distribution,
- the moisture content.

As a rule of thumb, the dust and chips have three times smaller volume density than the solid wood. A wet sawdust has around 300 kg/m^3 density but a dry and smaller dust only about 150–180 kg/m^3. Note that a loose bulk material is not very stable, its volume density easily increases due to settling in time.

1.2 Particle Size Distribution

One of the most important characteristics of the produced dust and chips is their size distribution. The size distribution is determined by screening using a series of screens of different mesh sizes. The highest mesh size is 2 mm and the smallest is 40 or 60 μm. If a more accurate size distribution under 60 μm is needed, laser measuring equipment may be used measuring only the fine portion of dust as the residue under the smallest mesh size.

The size distribution of dust and chips is belonging as an artificially produced distribution, to the class of asymmetric distributions and it can well be described as a logarithmic normal distribution.

The measurement results can be plotted either as a distribution function or a cumulated probability function using special probability net, Figs. 1.1 and 1.2.

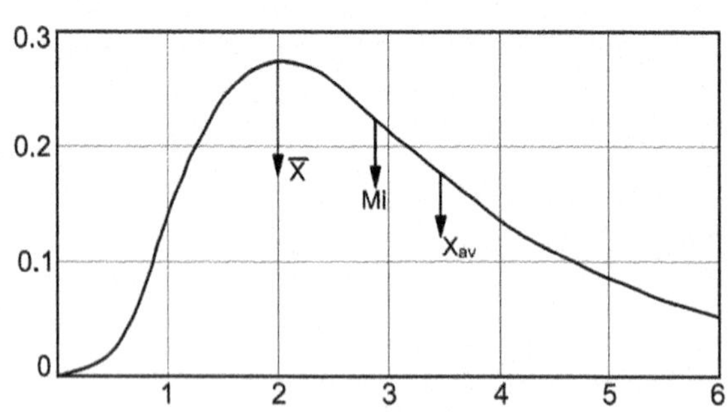

Fig. 1.1 Distribution function of particles. The mode $\overline{X} = 2$ mm, median Mi = 2.866 mm, geometric mean $X_{av} = 3.432$ mm, and $\sigma = 0.6$

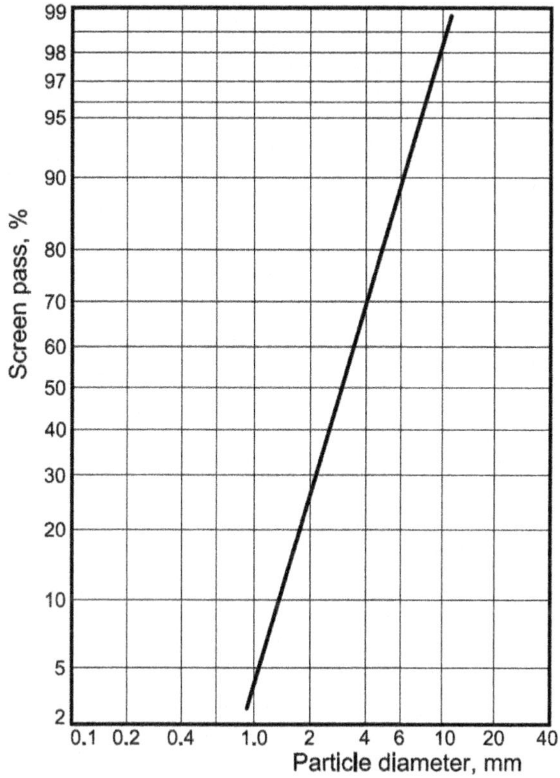

Fig. 1.2 Cumulated probability function of particles plotted in Fig. 1.1

The main characteristics of distribution are the mode, median, geometric mean, and the standard deviation.

The mode $\overline{X}$ is that particle size belonging to the maximum of distribution curve (Fig. 1.1). The median Mi separates the particles into two parts of the same quantity.

The distribution function is given by the following equation

$$y = \frac{1}{\sqrt{2\pi} \cdot \overline{X} \cdot \sigma \cdot e^{\frac{\sigma^2}{2}}} \cdot e^{-\frac{\left(\ln \frac{x}{\overline{X}}\right)^2}{2\sigma^2}} \tag{1.2}$$

Using the common sieve analysis, the experimental results are purposefully plotted in a probability diagram (Fig. 1.2). The two parameters of the distribution function can be determined in the following way. In Fig. 1.2 we read the corresponding particle diameters to 25, 50, and 75% probability and construct the quarter ratio

$$\mathrm{QR} = \frac{x_3 - M_i}{M_i - x_1} \tag{1.3}$$

where

x_3 is the diameter to 75%,
x_1 is the diameter to 25%.

The quarter ratio determines the standard deviation

$$QR = e^{0.675 \cdot \sigma} \tag{1.3a}$$

Similarly, we may read the particle diameters to the 10, 50, and 90% cumulated values and in this case

$$\frac{x(90) - M_i}{M_i - x(10)} = e^{1.28\sigma} \tag{1.3b}$$

and

$$\overline{X} = M_i \cdot e^{-\sigma^2} \tag{1.4}$$

The geometric mean X_{av} is integrated from Eq. (1.2) as

$$X_{av} = \frac{1}{\sqrt{2\pi} \cdot \overline{X} \cdot \sigma \cdot e^{\sigma^2/2}} \int_0^\infty x e^{-\left(\ln(x/\overline{X})\right)^2/2\sigma^2} dx \tag{1.5}$$

which yields also the following relations:

$$X_{av} = M_i \cdot e^{\frac{\sigma^2}{2}} = \overline{X} \cdot e^{1.5 \cdot \sigma^2} \tag{1.6}$$

Theory supposes that particle size varies between zero and infinity which means a full distribution. In practice we always have a maximum particle size x_{max} and, therefore, in Eq. (1.5) we should use this value as the upper limit of integration. The value of x_{max} may considerably influence the geometric mean which will be always smaller, due to the cut of given part of particles.

A detailed analysis has shown that the relative decrease in geometric mean $\left(X_{av}/X_{av}^{\infty}\right)$ depends on the x_{max}/X_{av}^{∞} ratio and the standard deviation. This relationship is plotted in Fig. 1.3 on half-logarithmic scale (Sitkei 1994).

Example. Concerning Fig. 1.2 and we take the cut-off size $x_{max} = 10$ mm. The volume or weight of particles up to 10 mm is 98.13% and the portion over 10 mm is 1.87%. The ratio $x_{max}/X_{av}^{\infty} = 2.914$ and the intersection with $\sigma = 0.6$ is at the line $X_{av}/X_{av}^{\infty} = 90\%$. Therefore, the expected value of the geometric mean is $X_{av} = 0.9 \times 3.432 = 3.09$ mm. If x_{max} is taken at 8 mm, then $\Delta Q = 4.36\%$, $x_{max}/X_{av}^{\infty} = 2.33$ and $X_{av} = 2.78$ mm.

In practice, the size distribution of dust particles is determined by sieves of different screen sizes. The narrow fractions of sieve residues are weighted and plotted either as a distribution curve or a cumulated probability curve (Figs. 1.1 and 1.2). As

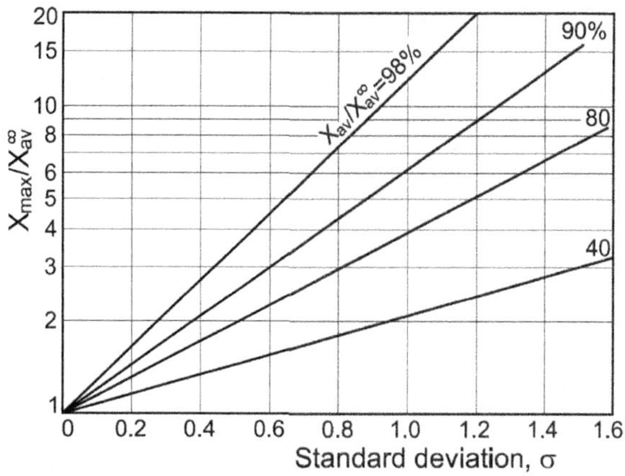

Fig. 1.3 Influence of cut-off size x_{max} on the decrease of geometric mean

a consequence, they are distributions related to mass. That is a reliable and convenient measurement method and, therefore, the limit values given in standards are also referred to mass units.

In some cases, such as health risks, the size distribution by number may be important. In other cases, such as the intensity of dust explosion, the specific surface area of a given particle distribution may be of interest. These particular cases will shortly be treated in Sects. 7.1 and 7.2.

1.3 Effect of Machining and Wood Species

Woodworking machines are generally working under different operational conditions such as cutting speed, width and depth of cut, feed speed, tool sharpness and geometry, and even wood species. All these factors influence the particle size distribution and the maximum particle diameter.

Figure 1.4 shows the wide range of distributions and the differences in the average particle size, for example, median, in the maximum particle diameter, and the fullness of distribution (Sitkei 1994). For example, in curve No. 7 some 6% volume is absent from the full distribution. Please note, that in the probability diagram the curves may lean on right or left side concerning sieve pass or sieve residue.

Fiberboard glued and pressed from fine particles gives finer chips compared to solid wood. The evaluation of distributions presented in Fig. 1.4 is summarized in Table 1.1. The high standard deviation and cut-off have resulted in a two and half-time decrease in the geometric mean.

The direction of cut (length cut or cross cut) has also some effect on the particle distribution. Especially conifers tend to produce slightly higher median at length cut, due to the oblong shape of particles (Fig. 1.5; Reisz and Magoss 2013). Planers may produce quite flaky chips giving a large amount of big particles. In this case, the

Fig. 1.4 Particle size probability curves for different woodworking machines. 1—Sanding, 2—circular saw (fibreboard), 3—bandsaw, 4—framesaw, 5—circular saw, 6—planing, 7—molder

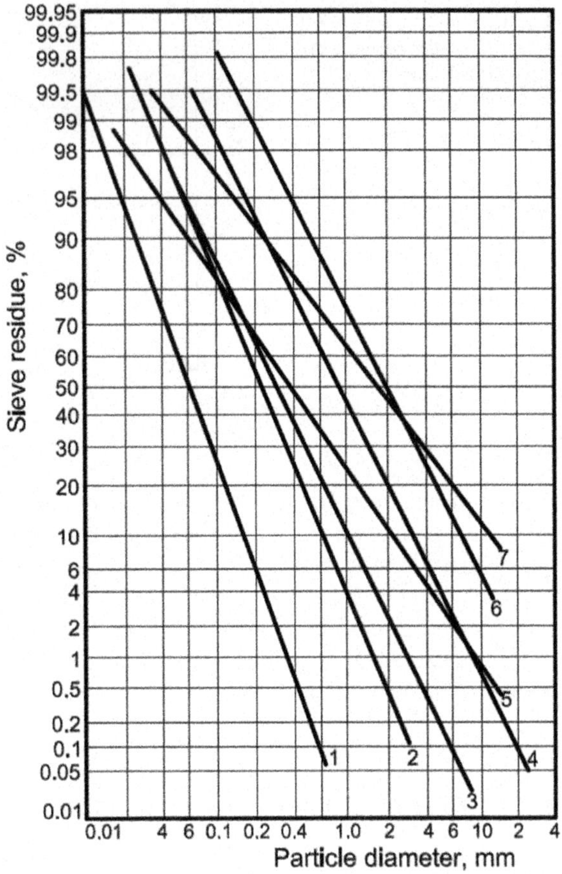

Table 1.1 The main characteristics of the distribution curves given in Fig 1.4

No.	Mi	QR	σ	$\overline{X} = \text{Mo}$	X_{av}^{∞}	X_{av}	$-\Delta Q\%$
1	0.0665	1.757	0.85	0.032	0.0954	0.091	0.47
2	0.23	1.9	0.97	0.0894	0.368	0.35	0.4
3	0.325	1.865	0.94	0.134	0.505	0.5	0.1
4	0.89	2.13	1.14	0.238	1.704	1.58	0.63
5	0.39	2.27	1.23	0.086	0.83	0.76	0.15
6	2.1	1.94	1.0	0.773	3.46	2.9	2.45
7	1.78	2.88	1.6	0.137	6.4	2.53	9.14

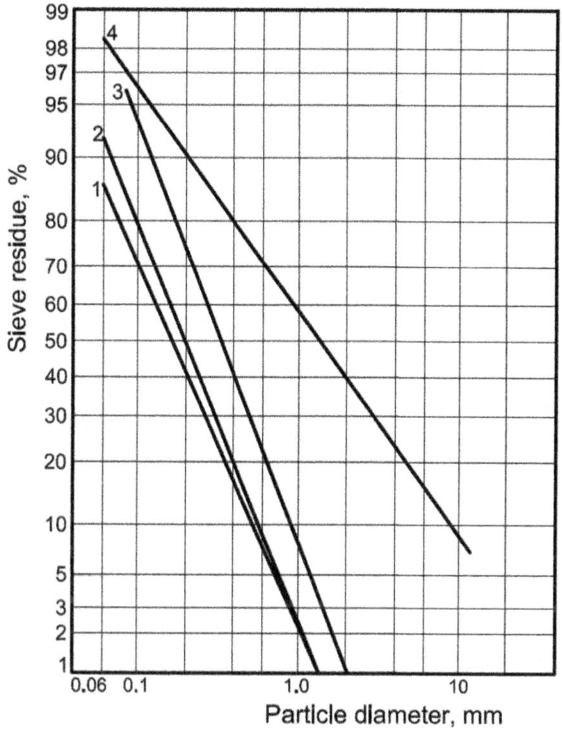

Fig. 1.5 Particle size distribution at length cut. 1—Beech, bandsaw, 2—oak, bandsaw, 3—pine, bandsaw, 4—pine, planer

distribution of particles does not follow any theoretical distribution in a sufficient manner. Small particles in the 5–40 μm range may show also some anomalies due to their tendency to coagulation and sticking to bigger particles. Figure 1.6 shows distribution curves for the fine fraction of different wood species. In the range of 4–40 μm a deviation can be seen in the course of the curves. Furthermore, even in the case of sieve pass, we have no sharp cut-off of particles larger than the mesh size; oblong particles get through the sieve, which may have larger size in perpendicular direction.

The moisture content of wood may have considerable effect on the particle size distribution. The higher moisture content increases the adhering of small particles to the bigger ones reduces the apparent number of small particles (see Figs. 3.6 and 5.16).

1.4 Friction Coefficient

The friction coefficient plays an important role in almost all technological processes. In the handling of dust and chips, the friction coefficient influences the energy requirement of transportation and the general behavior in bulk and flow.

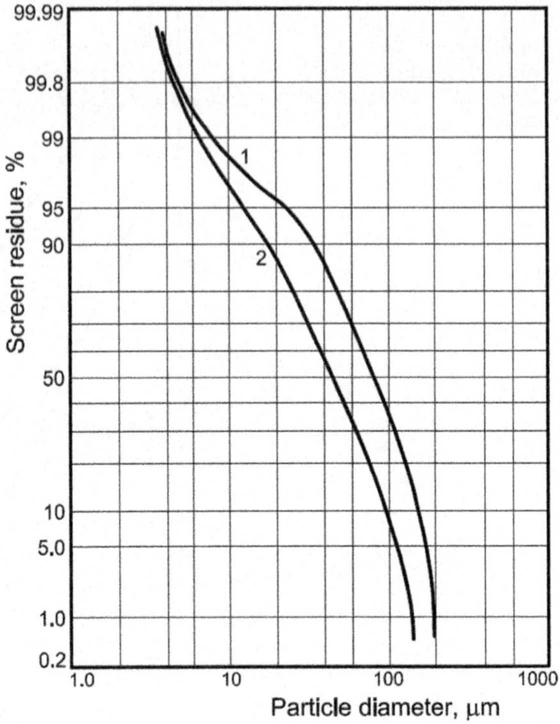

Fig. 1.6 Probability function of the fine fraction for different saw dusts (pine, beech, and oak). 1—Screen residue on 63 μm mesh size, 2—screen pass on 63 μm mesh size

The relationship between the force F required to start or maintain movement of a body with a weight G may be expressed by Coulomb's law in the simple form

$$F = \mu \cdot G$$

where μ is the static or dynamic friction coefficient.

The force required to initiate displacement is termed the static friction force.

In the above formulation is assumed that the friction force, or friction coefficient, is

- proportional to the normal force,
- independent of the dimensions of sliding surface,
- independent of the sliding velocity,
- dependent on the nature of contact surface.

In practical cases, the above assumptions are not always valid. Concerning biological materials, their contact surface under the action of pressure and heat may be modified whereby the friction coefficient may also vary.

Concerning wood chips, the most important influencing factors are the moisture content, wood species, and the surface properties. Figure 1.7 shows friction coefficients on steel plate for different wood species as a function of moisture content.

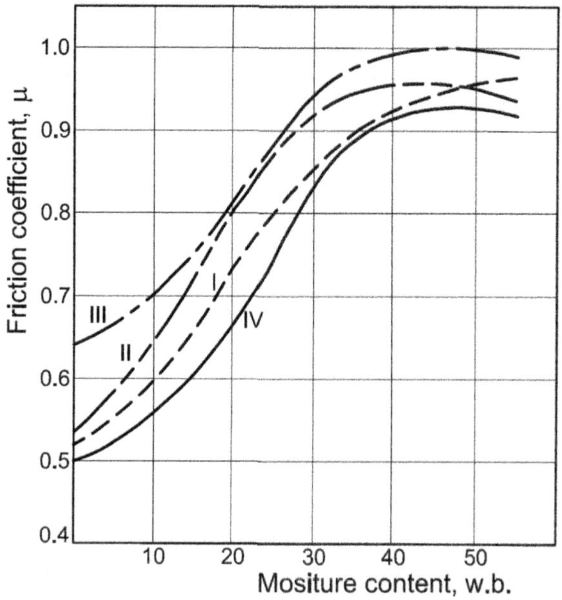

Fig. 1.7 Friction coefficient on steel plate. I—Cottonwood, II—pine, III—oak, IV—black locust

The moisture content of up to 40% has a fundamental effect on the friction coefficient. The wood species and size distribution have moderate effects on the friction coefficient.

A special feature of granular bulk materials is that the particles inside the heap may slide one on another. The friction coefficient between the particles is termed internal friction coefficient. The internal friction determines the shear stresses in bulk material and hereby also the flow properties. The stress in the shear plane depends on the loading normal stress and the internal friction coefficient

$$\tau = c + \sigma \cdot \tan \phi_i \tag{1.7}$$

where c means the cohesion depending on the consolidation of the granular material (Sitkei 1986).

The internal friction coefficient or its angle depends on the moisture content and wood species. Figure 1.8 demonstrates the variation of friction angle as a function of moisture content (Sitkei 1994). The moisture content is also here a definite influencing factor.

A bulk granular material, due to its own weight, may be compacted as a function of time (settling). A compacted material requires higher shear stress for starting to move along the shear plane due to its cohesion (see Eq. 1.7). Figure 1.9 shows experimental results with pine sawdust (frame saw) for differently consolidated (settled) samples, defining the initial apparent internal friction angle. For stationary sliding, the internal friction angle will be constant (Sitkei 1994).

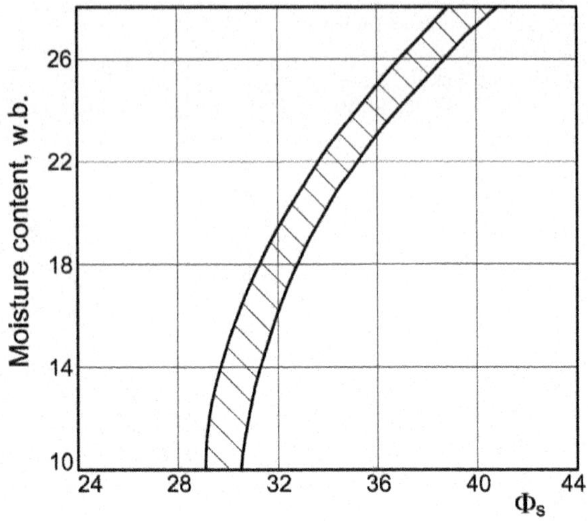

Fig. 1.8 Internal friction coefficient for pine wood

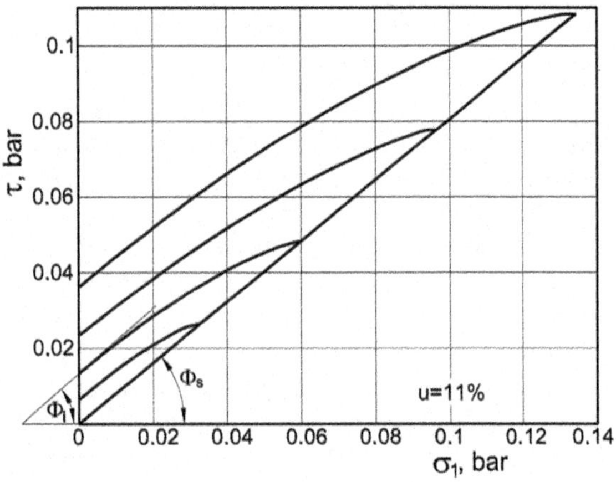

Fig. 1.9 Shear stress–normal stress relationship for differently consolidated pine sawdust and definition of an apparent initial internal friction angle ϕ_i

Important practical observation that sawdust that longer time is stored in a bin will flow, but with difficulties.

1.5 Side Pressure on Vertical Wall

In the theory of elasticity, it is well known that a column loaded vertically has also lateral strain which is given by the Poisson's ratio. In granular bulk materials stored in a bin, also lateral pressure occurs depending on several influencing factors. The ratio of the horizontal pressure p_h by the vertical pressure p_v is termed the lateral pressure coefficient

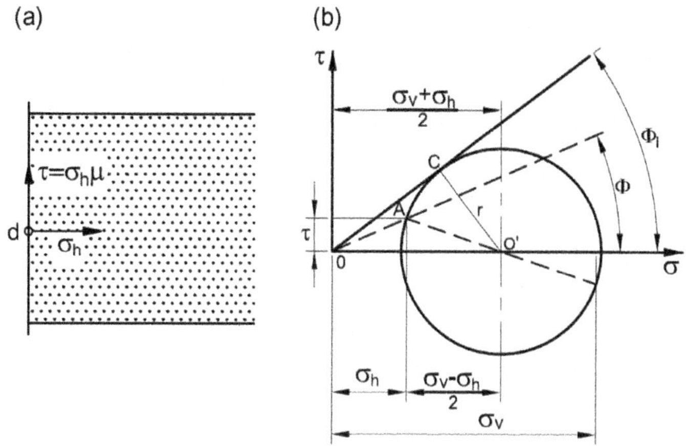

Fig. 1.10 Equilibrium of a granular medium along a wall

$$k = p_h / p_v$$

The lateral pressure coefficient is minimum when the greatest principal stress σ_1 acts vertically. Along a vertical wall shear stresses appear due to the settling of the material and, therefore, the greatest principal stress σ_1 encloses an angle Ψ with the wall (Sitkei 1986). The shear stress on the wall is simply given as

$$\tau = \mu \cdot \sigma_h = \sigma_h \cdot \tan\phi$$

where ϕ is the angle of friction ($\mu = \tan\phi$).

The equilibrium conditions are demonstrated in Fig. 1.10. The greater the angle of friction ϕ, the closer will point A come to point C on the Mohr circle, and the higher will be the value of the lateral pressure coefficient.

The lateral pressure coefficient along the wall may be calculated for an ideal bulk material (Fig. 1.10) from the following relationship (Zenkov 1964):

$$k = \frac{1}{1 + 2\mu_i^2 + 2 \cdot \sqrt{(1 + \mu_i) \cdot \left(\mu_i^2 - \mu^2\right)}} \tag{1.8}$$

In the case of $\mu_i = \mu$, the above expression simplifies to

$$k = \frac{1}{1 + 2\mu_i^2} \tag{1.8a}$$

Example: $\mu_i = 0.65$, $\mu = 0.6$ then Eq. (1.8) supplies $k = 0.402$. If we use the simple Eq. (1.8a), then $k = 0.542$. In practical cases, the disturbing factors are the time of consolidation in the storage bin and the non-uniform properties of dust and chips (size distribution, moisture content). Generally, the lateral pressure coefficient rarely exceeds the value of 0.55.

1.6 Terminal Velocity

A body placed into a flowing medium is subjected to a drag force consisting of frictional forces and a force due to asymmetrical pressure distribution on the body. This drag force is calculated from the equation

$$F_d = c_e \cdot A \cdot \frac{\rho}{2} \cdot v^2 \tag{1.9}$$

where

c_e is the drag coefficient,
A is the cross-section of the body normal to the flow,
v is the flow velocity relative to the body,
ρ is the air density.

Values of drag coefficient have been determined for bodies of regular shapes (sphere, cylinder, flat disk) and are generally plotted on logarithmic coordinates as a function of Reynolds number *Re*.

Sadly, wood dust particles are of irregular shape with rough surface and their drag coefficients depend on the instantaneous position (orientation). We can put, however, some upper and lower limit values based on regular shapes. The sphere has a value of $c_e = 0.44$ and the flat disk $c_e = 1.1$. Dust particles with rough surfaces come closer to the flat disk and their drag coefficients may vary between 0.7 and 0.9.

A free-falling body, after a given time, falls with a constant velocity which is termed the terminal velocity. In this case, the force due to air resistance is identical to the weight of the body:

$$c_e \cdot A \cdot \frac{\rho}{2} \cdot v^2 = G$$

This equation supposing a spherical body with diameter d and a volume weight γ_m, can easily be solved for the terminal velocity

$$v_{\mathrm{cr}} = \sqrt{\frac{4d \cdot \gamma_m}{3c_e \cdot \rho}} \tag{1.10}$$

If we use $c_e = 0.8$ and $\rho = 1.205$ (20 °C), then the above equation has the form

$$v_{\mathrm{cr}} = 1.176\sqrt{d \cdot \gamma_m} \tag{1.10a}$$

In laminar flow the drag coefficient is $c_e = 24/Re$ and, therefore, the drag force has the following expression

$$F_d = 3\pi d\eta v \tag{1.9a}$$

which equals the particle weight. The terminal velocity is determined by the equation

$$v_{\mathrm{cr}} = \frac{d^2\gamma_m}{18\eta} \tag{1.10b}$$

where

γ_m is the specific weight of particles, N/m^3.
η is the dynamic viscosity of air, Ns/m^2.

In the practice, this equation for irregularly shaped particles gives very approximate values. It is difficult to calculate an equivalent diameter and, for a small particle, the volume weight is not a simple task. Due to the voids (vessel and tracheids), a small and irregularly shaped particle may have smaller apparent density related to the cross-section which is effective for the drag force. Therefore, an equivalent particle diameter for effective cross-section and weight is not identical.

The experimental determination of terminal velocity is also not a simple task. The handling of single particles is difficult and we can rather measure a narrow fraction produced by sieves.

Figure 1.11 shows experimental results of terminal velocity for the particle diameter range of 40–1000 μm. Three wood species have been used with narrow size fractions. The theoretical values were calculated for spherical particles with drag coefficient of 0.8. Particles down to 100 μm follow well the theoretical curves but

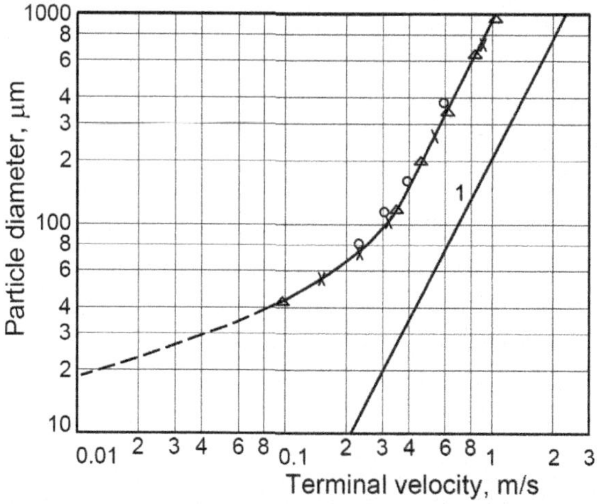

Fig. 1.11 Terminal velocities measured for narrow dust fraction of pine, beech and oak species 1–500 kg/m^3

shifted toward smaller densities due to irregular shape, lower apparent density, and higher drag coefficient. Under 100 μm the terminal velocity rapidly decreases and a particle around 5 μm may be floating for longer time in a resting air.

The large discrepancy between calculation and measurement can be explained by the fact that concerning particles with irregular shapes and very rough surfaces with voids, the equivalent diameter of the blown surface and that of the mass may considerably deviate from one another. That means that we over-estimate the weight of particles. If we use different equivalent diameters for blown surface d_A and particle weight d_m, and take a higher drag coefficient c_{er} (0.8–0.9), we can rewrite the common terminal velocity equation in the form:

$$d_m = k_1 \cdot d_0 \quad \text{and} \quad d_A = k_2 \cdot d_0$$

$$v_{cr}^{+} = \sqrt{\frac{k_1^3}{k_2^2}} \cdot \sqrt{\frac{4 \cdot d_0 \cdot \gamma_m}{3 \cdot \rho_0 \cdot c_e}} = \sqrt{\frac{k_1^3}{k_2^2}} \cdot v_{cr} \tag{1.10c}$$

where d_0 is the particle diameter obtained from screen measurement.

Sadly, d_m and d_A cannot be determined with direct methods. Using measured terminal velocities, they can be determined with the assumption that the particle diameter obtained from screen measurements more or less equals the equivalent blown diameter d_A. With this assumption, as a first estimate, the ratio d_m/d_A is around 2/3 for particles over 100 μm.

It is very important, however, that the measured terminal velocities for particle sizes over 100 μm follow the turbulent drag law, despite the low particle *Re*-numbers not more than 2. Reason is for it that the main airflow is definitely turbulent. For example, in a cyclone, we have *Re*-numbers for the airflow around 10^5 but the particle *Re*-number for 20 μm diameter is not more than 0.1! The development of drag force fundamentally depends on the turbulence level of air stream.

Particles below 100 μm show further increased deviations due to complicated hydrodynamic phenomena and increased influence of particle shape morphology. The influence of irregular shape and hairy surface further increases and the equivalent effective diameter for mass and blown surface further deviate from each other. If particles move in a turbulent airflow then the turbulent velocity fluctuation (large-scale turbulence) has a definite interaction with small particles.

The experimental procedure for determination of terminal velocity of small particles with appropriate level of turbulence is also not a simple task. We need very low flow velocity with high intensity of turbulence which can be realized by using special turbulizator.

It is important to note that the terminal velocity is more determinative for the selection of minimum air velocity at transporting granular materials with impact losses (see in Sect. 4.4.1) and much less with friction losses.

The free fall of particles in resting air is a common problem. In principle, we may have turbulent and laminar motion of particles depending on their diameter and terminal velocity. The general equation of motion reads

$$m\frac{\mathrm{d}v}{\mathrm{d}t} + F_d - mg = 0$$

where the aerodynamic force F_d is given by Eq. (1.9) for turbulent and Eq. (1.9a) for laminar motion. The terminal velocity is given by Eqs. (1.10) and (1.10a), respectively.

Concerning turbulent motion, the solution of the above differential equation yields the displacement of the particle in the following form (Sitkei 2012):

$$S = \frac{v_{\mathrm{cr}}^2}{g} \ln \cosh(t/T)$$

where $T = v_{\mathrm{cr}}/g$.

If $t = T$, then the displacement is $0.434.v_{\mathrm{cr}}^2/g$ which is a characteristic value for the distance of acceleration. Over $t/T = 2$ the acceleration ceases and the particle falls with its constant terminal velocity.

In practice, the laminar motion is more frequently encountered. In this case, the aerodynamic force is given by Eq. (1.9a) and integration of differential equation yields.

$$v = v_{\mathrm{cr}}\left(1 - \mathrm{e}^{-t/T}\right)$$

and

$$S = v_{\mathrm{cr}}t - v_{\mathrm{cr}}T\left(1 - \mathrm{e}^{-t/T}\right) \cong v_{\mathrm{cr}}t$$

In this case, the value of T is small and, therefore, the particle is falling with the constant terminal velocity.

Example: A dust particle of 100 μm has a terminal velocity of 0.3 m/s and $T =$ 0.0306 s. Taking $t/T = 2$, the displacement is 10.4 mm. After this displacement, the particle is settled with constant velocity v_{cr}.

1.7 Particle Motion in Air Stream

As we have seen in previous section, the aerodynamic drag force depends on the relative air stream velocity between the particle and air. In a horizontal tube, a given relative velocity is necessary to bring the particle into a floating position. A resting particle will be flowing around by the air and, against the own weight, a lifting force develops which brings the particle into floating state (Fig. 1.12a). The floating particle starts to move with the air with increasing relative velocity v_m/v_0. Depending on the terminal velocity of particle, it reaches its steady-state motion condition at a given airflow velocity v_0^*. That is the minimum air velocity that makes the conveying of a particle possible (Fig. 1.12b).

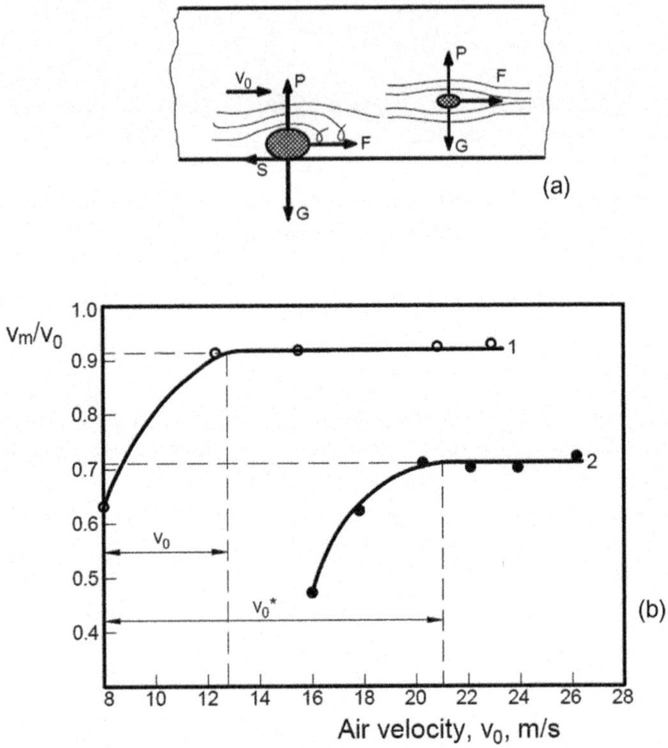

Fig. 1.12 Forces acting on a particle in horizontal tube (**a**), development of steady motion of particles as a function of air velocity (**b**). 1—Match-sticks, 2—high-density big chips

In practice, we always have a particle size distribution and the biggest particles, their density, and also the transport concentration influence the necessary air velocity for clogging-free operation. A more detailed discussion on the necessary air velocity will be given in Chap. 4 (Tauber 1978).

References

L. Reisz, E. Magoss, Particle size distribution for different wood species and woodworking operations. Derevoobrabatyvayushchaya Promyshlennost (1), 38–45 (2013). ISSN: 0011-9008

G. Sitkei, *Mechanics of Agricultural Materials* (Elsevier, Amsterdam, 1986), pp. 284–402

G. Sitkei, *Faipari műveletek elmélete* (*Theory of Wood Processing*) (Budapest, 1994), pp. 485–507

G. Sitkei, Particle motion with air resistance. Prog. Agric. Eng. Sci. 1–15 (2012)

Б. Таубер, *Внутризаводский транспорт* (*Internal Transport in Wood Industry*) (Машгиз, Москва, 1978)

Р. Зенков, *Механика насыпных грузов* (*Mechanics of Bulk Loads*) (Машгиз, Москва, 1964), p. 252

Chapter 2
Behavior of Dust and Chips in Bulk

As outlined in Sect. 1.4, granular bulk materials, due to their own weight, may be settled in time. As a consequence, their porosity decreases which modifies the shear stress-normal stress relationship and hereby the flow properties (see Fig. 1.9). These evidences led to the recognition that, in the case of granular bulk materials, it is purposeful to include the porosity as a variable into the stress state analysis. It is especially true if the particle can easily be deformed under relatively low loads. Fein wooden particles are belonging to this case.

In this chapter, we treat the state diagram and the stress state of loose granular materials, the flow properties, and their characterization. The compaction of bulk dust is treated only in the low-pressure range with worked examples for storage bins.

2.1 State Diagram of Loose Granular Bulk Materials

The porosity is an important physical property of porous and granular bulk materials. The porosity is calculated as the ratio of voids to the volume of the material. In the case of wooden particles, we distinguish internal and external pores. Under vertical load, in the first stage of volume reduction, the outer pores are consumed, and gradually the volume of inner pores will also be reduced (Kocsis and Csanády 2019). The reduction of inner pores depends on the hardness of particles, but the size of particles is also an influencing factor. Fein particles show a smaller apparent hardness than bigger particles of the same wood species.

Figure 2.1 illustrates the state diagram characteristic for granular bulk materials. The curve n_1–n_2 lies in the σ–ε plane and describes the initial compression of the material. Material compressed by the stress σ_1 is unloaded along the curve section *EF*G. Cohesive materials can also take up a certain tensile stress, whose limiting value is shown by the curve h_1–h_2 (Jenike 1964).

E. Magoss et al., *Dust Extraction and Handling in the Wood Industry*,
SpringerBriefs in Applied Sciences and Technology,
https://doi.org/10.1007/978-3-031-08915-2_2

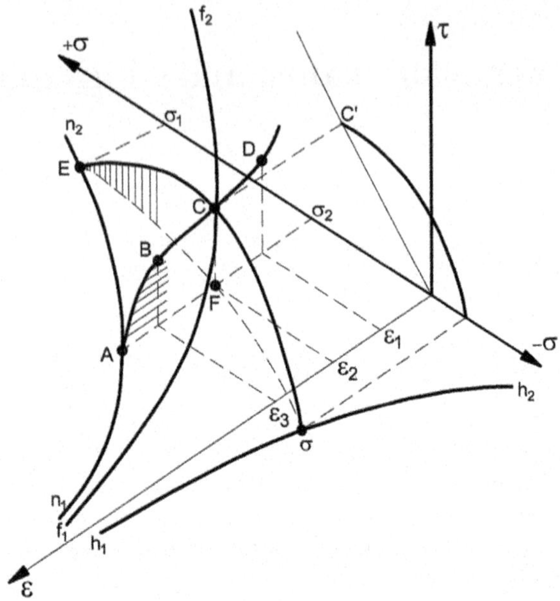

Fig. 2.1 State diagram of granular bulk materials

The τ–σ–ε relationship corresponding to the stationary flow state is described by the three-dimensional curve f_1–f_2. Accordingly, the stationary flow state may be maintained under a normal stress σ_2 in the case of any preliminary consolidation by values of τ and ε corresponding to point *C*. In the case of higher initial pore volume ratio (ε_3), the material first assumes the stress τ corresponding to point *B*, and then at the start of flow is compressed and becomes capable of assuming a higher stress τ. Thus, the material again reaches the state corresponding to point *C*. For greater preliminary consolidation (ε_1), the material assumes the stress corresponding to point *D*; at the start of flow, it loosens and again reaches the state corresponding to point *C*. However, in the latter case, a higher stress τ is required to initiate flow than to maintain it.

The curves n_1–n_2, f_1–f_2, and h_1–h_2 supply a three-dimensional surface whose sections are the curves *ABCD* and *ECG*. For state characteristics below this surface, only elastic deformation occurs; state characteristics above this surface are not possible. Flow of material is possible for state characteristics found along the surface.

The projection of the curve f_1–f_2 and τ–σ plane is approximately a straight line passing through the origin. The projection of the curve section *CG* found above these straight-line supplies in a given case the stress τ required to initiate flow of material consolidated by normal stress σ_1 in the interval 0–σ_2.

Materials inclined to settling will consolidate over time to a low pore volume, i.e., they become over-consolidated relative to the given normal stress. A high-stress τ is then required to initiate flow. Media resting in bins can flow out only if the stresses τ due to weight forces attain the critical values required to initiate flow.

2.2 Stress State of Granular Bulk Materials

Granular bulk materials may be treated, with certain restrictions, as continuous solid materials. If the size of individual grains is small compared to the volume of the pile and the grains may be regarded as elastic (i.e., without the appearance of stresses σ and τ sufficient to cause plastic deformation), and provided that the mechanical properties are identical in every direction, then the stresses τ and σ caused by a load may be interpreted according to the relation

$$\tau = \tau_0 + \sigma_i \cdot \tan \phi_i$$

Characteristic form of this function is shown in Fig. 2.2. The forces acting on a pile may be of two types: internal farces acting on each individual grains (weight and inertial forces), and external forces acting on the surface of the pile. The bulk material is in equilibrium when there is no displacement among individual grains.

To determine the plane stress state of a triangular prism, on whose orthogonal sides the principal σ_1 and σ_2 are acting, is sampled from the bulk body. On the third side, lying under the angle α_1, the stresses σ and τ which arise may be calculated from the equations (Zenkov 1964):

$$\sigma = \sigma_1 \cdot \cos^2 \alpha_1 + \sigma_2 \cdot \sin^2 \alpha_1 \tag{2.1}$$

and

$$\tau = [(\sigma_1 - \sigma_2)/2] \cdot \sin^2 \alpha_1 \tag{2.2}$$

The stresses σ and τ may also be determined graphically by means of Mohr's circle. Knowing the principal stresses σ_1 and σ_2, a circle may be drawn in the τ–σ

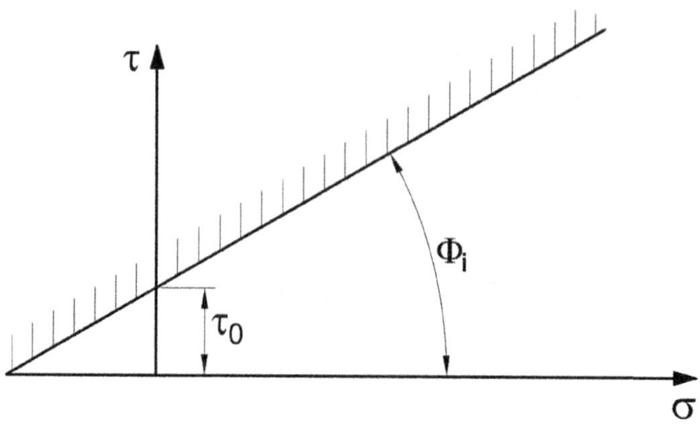

Fig. 2.2 The shear and normal stress relationship

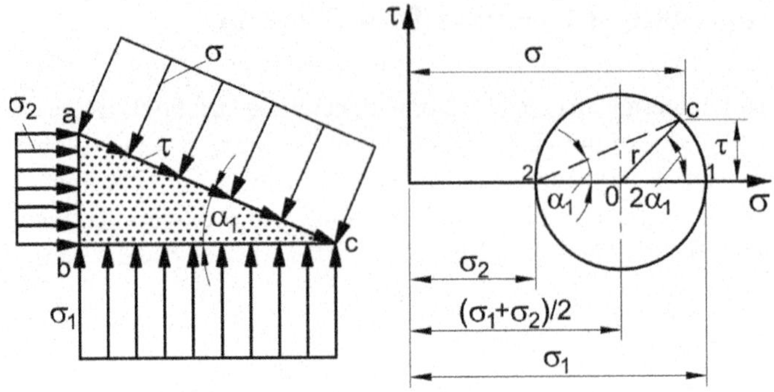

Fig. 2.3 Mohr circle for derivation of the stress state of granular pile

system of coordinates. The straight line is drawn at an angle α_1 from point 2, or at an angle $2\alpha_1$ from the center, yields the point C as the intercept on the circumference, whose coordinates give the required stresses σ and τ (Fig. 2.3).

Bulk material, as a medium, will be in equilibrium if the relationship between the stresses τ and σ described by Eq. (1.7) prevails. Accordingly, the Mohr circle of stresses in Fig. 2.2 is always on the inside of the envelope curve or contacts it from the inside. This case is termed the equilibrium limit position (Sitkei 1986).

Bulk material is generally surrounded by walls on several sides, while its upper surface is free. On the free surface, no stresses arise, while along the wall a friction stress is given by

$$\tau = \mu \cdot \sigma_h$$

arises, where μ is the friction coefficient between the bulk material and the wall. There may be cases, where the medium is in equilibrium (grains do not move in relation to each other) but the whole body slides along the limiting walls.

When the equilibrium state for the bulk medium ceases to exist, the individual grains slide over each other along elemental planes, enclosing an angle with the plane of the local principal stress. The elemental slip planes supply the slip surface.

2.3 Flow Properties of Bulk Materials

Temporary storage of bulk dust and chips often occurs in silos. The flow of granular material from circular or square orifice of silos plays an important role in numerous technological processes (discharge of silos, flow from bins into dosage elements, etc.). The outflow cannot be calculated with the generally known hydrodynamic relationships, since the flow properties of granular materials differ fundamentally from those of liquids.

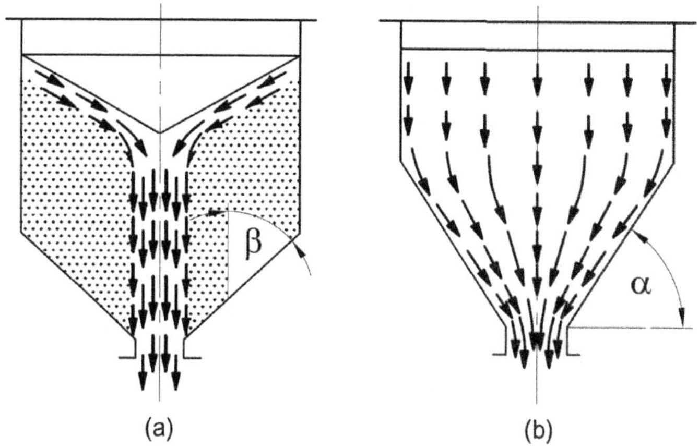

Fig. 2.4 Typical outflow patterns from bins: funnel flow (**a**) and mass flow (**b**)

Typical outflow modes are shown in Fig. 2.4. When the base plane of hopper encloses a relatively small angle with the horizontal, so-called funnel-type outflow occurs. The material moves only in the column located above the orifice and flows into the column from the upper layers. Another typical form of outflow is the mass flow. In this case, the whole volume moves and accordingly inflow occurs from the whole bin. Which mode of outflow occurs depends primarily on the hopper slope angle and the friction coefficient on the wall (Sitkei 1986).

Flow from bins is usually disturbed by the phenomenon of arching when the material remains in the form of an arch and leans against the wall. Arching remains stable if the strength of material σ_c at the point of support exceeds the compression stress σ_b originated from mass loads. The arch has a free surface on which the smaller principal stress is zero and, therefore, the strength σ_c is determined by the Mohr circle going through the origin ($\sigma_z = 0$) and contacts the appropriate yield locus, Fig. 2.5 (Jenike 1964). Figure 2.5 is obtained in the following way. The maximum shear stress

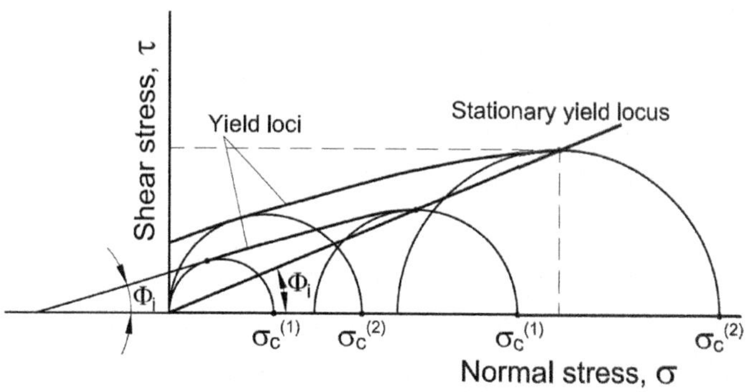

Fig. 2.5 Initial yield locus and stationary yield locus for granular materials

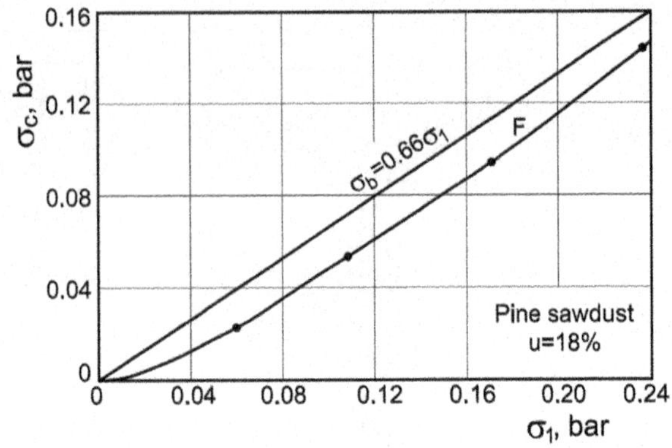

Fig. 2.6 The σ_c–σ_1 relationship for pine sawdust

is determined as a function of normal stress using a shearing box, applying a given consolidation stress σ_1. By applying several consolidations stresses σ_1 the straight line of the stationary yield locus and the curves of the yield loci corresponding to individual σ_1 values may be plotted. It may be seen from the Mohr circles drawn in the figure that for a consolidation stress σ_1 the development of stationary flow will be ensured by a normal stress σ corresponding to the center of the Mohr circle, without the occurrence of any peak value of shear stress. For lower normal stresses, the material is over-compacted, and so a higher stress τ is needed to initiate flow; the curves of the yield loci lie above the straight line. Mohr circles are also constructed through the origin ($\sigma_z = 0$) for the curves of the yield loci, and the main stress σ_c obtained in this way is characteristic of the strength of cohesive material. The ratio of the stresses σ_c and σ_1 is characteristic of the flow properties of bulk granular materials, and the relationship obtained is termed the characteristic flow function (*F*) (Jenike 1964).

As an example, Fig. 2.6 shows an experimentally determined σ_c–σ_1 relationship (Sitkei 1994). Outflow is possible if σ_c is smaller than the also experimentally determined σ_b function for a given hopper and granular material. Figure 2.6 shows a case when the outflow is possible whatever the stress state (σ_c is always less than σ_b).

Granular materials may be classified according to the ratio σ_1/σ_c as follows:

$\sigma_1/\sigma_c < 2$, highly cohesive materials and the outflow from a bin can be ensured only by auxiliary equipment;
$4 > \sigma_1/\sigma_c > 2$, cohesive materials;
$10 > \sigma_1/\sigma_c > 4$, easy-flowing materials;
$\sigma_1/\sigma_c > 10$, free-flowing materials.

Figure 2.5 shows that the investigated pine sawdust has σ_1/σ_c values between 1.7 and 2.0 and belonging to the highly cohesive material class. In the practice, therefore, flow aid devices are often used. Installation of flow aid device means additional cost but it may also reduce the overall cost of storage equipment. The free outflow of the material requires a conical bottom section which, in the case of small cone angle, considerably increases the height of the silo and also its cost. The use of a flow aid device enables an almost flat bottomed silo with much lower height.

2.4 Compaction of Bulk Material

Compaction of granular bulk materials under external load plays also an important role in many technological processes. In resting granular bodies stored in silos the load is originated from the own weight. In this case, the load is more or less proportional to the height and volume weight of the material. Taking a 5 m high silo, the vertical pressure in the bottom part rarely exceeds the 1×10^4 N/m^2 = 0.1 bar value. Nevertheless, considerable settling (consolidation) may occur as a function of storage height, which fundamentally alters the flow properties (see in previous Section).

In order to reduce the volume of loose granular materials (briquette and pellet making), high pressures are used which may have values up to 1500 bar. Here we do not treat high-pressure compaction and readers interested in this topic are referred to special literature (Kocsis and Csanády 2019).

Vertical pressure in a bulk material may be calculated from the simple relationship

$$\Delta p = \gamma_m \cdot h \quad \mathrm{N/m^2} \tag{2.3}$$

where

γ_m is the volume weight, N/m^3,
h is the height of the material.

Relationship between the strain (deformation) and volume density is given as

$$\rho = \frac{\rho_0}{1 - \varepsilon} \tag{2.4}$$

where ρ_0 is the initial density of the loose dust. Concerning a wood species with solid density around 500 kg/m^3, the expected loose chip density is one-third of the solid density, and it is about 160 kg/m^3.

Figure 2.7 shows the compaction of dry sawdust in a special plot (Sitkei 1994) where the apparent elastic moduli E can also be read. The relationship is described by the following equation valid only for the given pressure range

$$p = 84\left(\frac{\varepsilon}{1 - \varepsilon}\right)^{2.8} \quad \mathrm{N/cm^2}$$

In order to obtain explicit expressions for the further discussion, we make some simplifications in the above equation:

$$\left(\frac{\varepsilon}{1 - \varepsilon}\right)^{n} \cong \varepsilon^{c \cdot n}$$

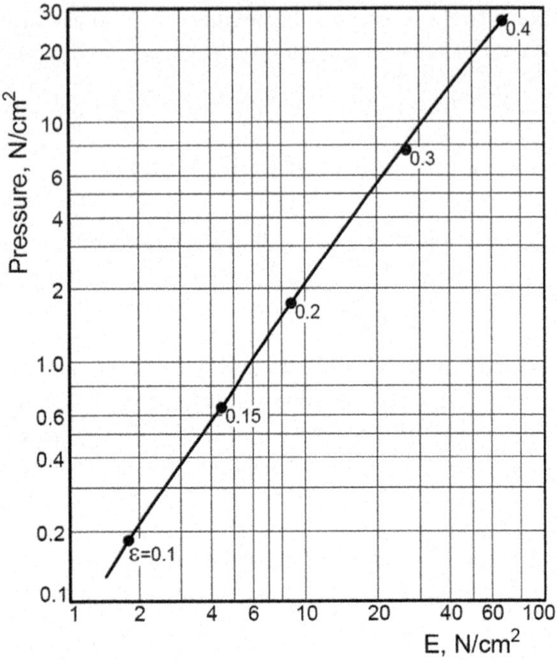

Fig. 2.7 The compaction of dry sawdust in the low-pressure range

Taking a 5 m high silo, the maximum static pressure does not exceed the value of 1 N/cm^2 and in this case, $c = 0.95$ is appropriate. The pressure-strain equation assumes the simplified form:

$$p = 84 \cdot \varepsilon^{2.66} \tag{2.5}$$

Note that this correction for higher pressure range is not valid.

Using Eqs. (2.3), (2.4), and (2.5) we may write

$$\varepsilon = \left(\frac{h \cdot \gamma_m}{84 \times 10^4}\right)^{0.376} = 1 - \frac{\rho_m}{\rho}$$

Taking $\gamma_m = 1600$ N/m^3, the above equation yields

$$\rho = \frac{\rho_0}{1 - 0.0949 \cdot h^{0.376}} \quad \mathrm{kg/m^3} \tag{2.6}$$

which describes the initial storage density as a function of filling height just after filling of a silo. As in viscoelastic bodies under constant load, a further slow creeping of the material (settling) takes place as a function of elapsed time.

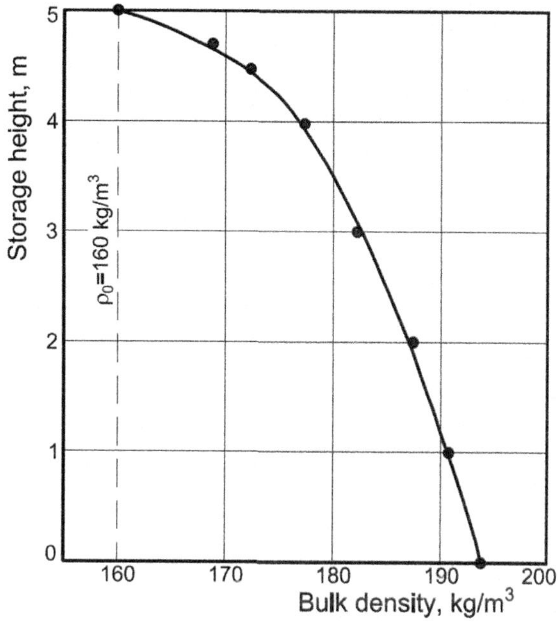

Fig. 2.8 Distribution of bulk density after filling in a silo of 5 m height

The distribution of volume density in a silo just after filling is shown in Fig. 2.8. In the bottom of silo the volume density noticeably increases which makes the outflow more difficult. During long time storage, the slow creep of the material causes an over-consolidation requiring higher shear stress to start flow in the bulk body. In practice a long time storage in a silo is not very common and, therefore, the compaction due to gravity should be taken into account.

In slender silos, where the height is greater than the diameter, in the vicinity of wall the vertical pressure decreases due to friction forces on the wall. This circumstance not much modifies, however, the vertical pressure in the axis of silo where the outflow orifice is placed.

References

A. Jenike, Storage and flow of solids. University of Utah, Engineering Experiment Station Bulletin, No. 123, S. 68–122 (1964)
Z. Kocsis, E. Csanády, *Theory and Practice of Wood Pellet Production* (Springer, Berlin, 2019)
G. Sitkei, *Mechanics of Agricultural Materials* (Elsevier, Amsterdam, 1986), pp. 284–402
G. Sitkei, *Faipari műveletek elmélete* (*Theory of Wood Processing*) (Budapest, 1994), pp. 485–507
Р. Зенков, *Механика насыпных грузов* (*Mechanics of Bulk Loads*) (Машгиз, Москва, 1964), p. 252

Chapter 3
Dust and Chip Collection

The collection of dust is a very important task that directly influences the work-room pollution and energy requirement of dust extraction. Complicated aerodynamic phenomena are taking place which can theoretically be followed with difficulties. As a consequence, practical solutions of higher efficiency are appeared only in the last time as a result of intensive research works.

3.1 Aerodynamics of Suction Devices

The chips separated by a rotating tool leave the edge approximately with the circumferential velocity of the tool. The tool edge acts similarly to a throwing disc (Sitkei 1986). The circumferential velocity is in the range of 20–50 m/s while the terminal velocity of particles depends mainly on the diameter and density of particles. The expected terminal velocity for diameters of 0.01, 0.1 and 1.0 mm is 0.2, 0.8 and 2.5 m/s, respectively. For smaller particles, the initial throwing velocity rapidly decreases within a short traveled distance.

The motion behavior of thrown particles in resting air can simply be estimated in the following way. As a first approximation, we neglect the effect of gravity and in this case, the equation of motion reads

$$m\frac{\mathrm{d}v}{\mathrm{d}t} = c_e \cdot A \cdot \frac{\rho}{2} \cdot v^2$$

or express the mass and cross-section with the diameter of particle, and remembering Eq. (1.10), yields

$$\frac{\mathrm{d}v}{\mathrm{d}t} = \frac{g}{v_{\mathrm{cr}}^2} \cdot v^2 = B \cdot v^2$$

E. Magoss et al., *Dust Extraction and Handling in the Wood Industry*,
SpringerBriefs in Applied Sciences and Technology,
https://doi.org/10.1007/978-3-031-08915-2_3

and further

$$\frac{dv}{v^2} = B \cdot dt$$

An integration, with the initial condition $t = 0 \rightarrow v = v_0$, yields the following expression (Sitkei 2012):

$$v_x = \frac{v_0}{v_0 \cdot B \cdot t + 1}$$

$$s_x = \frac{1}{B} \cdot \ln(v_0 \cdot B \cdot t + 1) \quad (3.1)$$

and

$$v_x = v_0 \cdot e^{-B \cdot S_x} \quad (3.2)$$

where

S_x is the traveled distance,
t is the elapsed time.

Equation (3.2) is plotted in Fig. 3.1. A small particle with 10 μm diameter loses its initial velocity of 50 m/s within 10 mm displacement. A particle with 100 μm dia. loses its velocity within 150–180 mm, while bigger particles will travel much longer distances. As a consequence, small particles require additional air motion to get safely them into the collecting tube.

In order to capture the chips thrown by the tool edge, it is purposeful to place the collecting tube such that the flying chips directly get into the tube, Fig. 3.2 shows a circular saw where the lower part of the saw is covered with a collecting tube receiving the thrown particles. The air motion within the housing induced by

Fig. 3.1 Relationship between throwing velocity and throwing distance for two different initial velocities and terminal velocities. $v_{cr} = 2.5$ m/s corresponds to $d = 1.0$ mm and 0.8 m/s to 0.1 mm

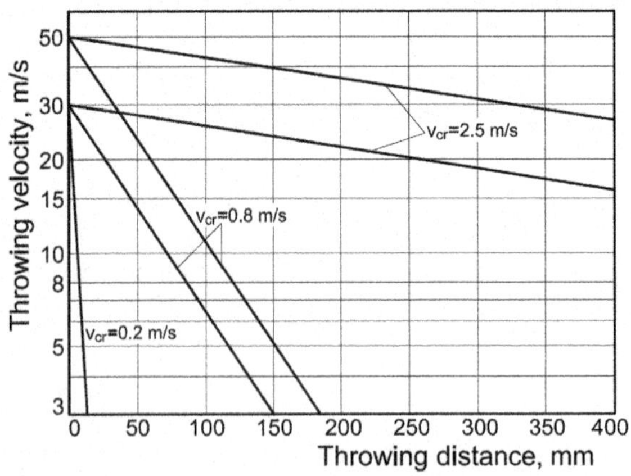

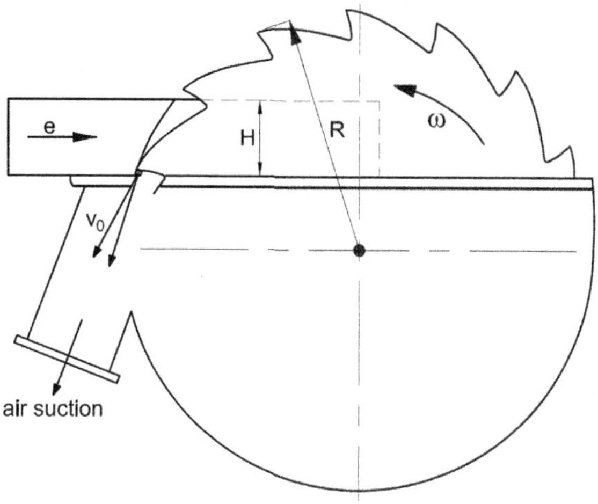

Fig. 3.2 Throwing of particles by the tool edge

the rotating tool promotes the transport of particles in a certain extent. Mounting a collecting sackcloth on the end of collecting tube, most of chips can be collected without additional air suction. This simple solution is applicable for intermittent operation only.

In continuous machining operations, an air suction is used which ensures a better capture of particles and their transportation to a storage bin. The air due to suction enters the suction tube through narrow gaps around the tool with high velocity and prevents the get out of fine particles into the surrounding environment and hereby air pollution at the workplace.

The placing and construction of suction hoods depend on the tool and operation design. In CNC machines the cutter head is moving and requires more open cross-section around it. As a consequence, the required specific air volume (m^3/h) may be higher. The required minimum air volume is generally prescribed by the manufacturers based on experiments.

3.2 Dust Collectors on Woodworking Machines

In the last century, the design of the suction hoods was improved significantly. The early woodworking machines did not contain any dust extraction device. In the first stage of the woodworking machines, the suction hoods usually were not placed in the proper position concerning the aim to catching more wood particles. Later, if the machine construction allowed the extraction hoods are designed and positioned according to the aerodynamic requirements.

In the last decades, the different types of ventilation tools appeared too, which are able to drive the wood particles into the inlet of the extraction pipe. There are

Fig. 3.3 Structure of the integrated tool (Gruing Aerotech)

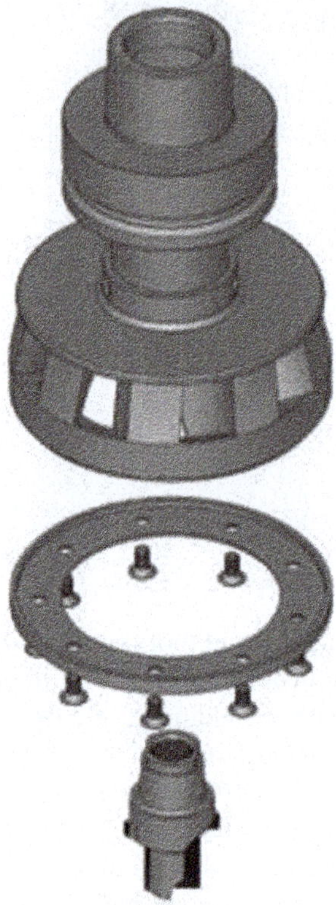

integrated tools with propeller, Fig. 3.3, and there are special tools whose construction is able to fulfill the cutting and the ventilation function too.

For example, the Leuco "I-System" (Fig. 3.4) is a specially designed milling tool with interior chip removal, which is used on some edge banding machines. These tools guide the particles inside, therefore the chips can thus be transported toward to exhaust hood more effectively than with normal tools.

Applying these types of woodcutting tools, the problem of the aerodynamically wrong-positioned suction hoods could be solved.

Fig. 3.4 Specially designed milling tool Leuco I-system

The design of proper extraction hood system for the CNC machines is a different task. In the previous section mentioned that the more open cross-section requires more air volume and, therefore, in order to decrease the plus energy consumption and to increase the efficiency of the extraction system some new inventions are applying in the practice nowadays. The theory behind these solutions is to utilize the motion of the recirculated air. Figure 3.5 shows a possible setup of the modified suction hood. The dusted air leaves the suction hood in axial direction, the most of the air is introduced through two blow-pipes in radial direction which is more effective to guide the particles into the sucked air stream. It is important, however, to reduce the "false" air through the slits between the hood edge and workpiece or machine table surface. If workpieces with varying shapes and sizes are machined, an optimal universal solution is rarely found.

It was an interesting observation that the distributions of oak dust extracted from a CNC machine are quite different if the moisture content is changed considerably. Figure 3.6 shows the experimental results for 10% and 18% moisture contents. At higher moisture content the fein fraction decreases due to particle coagulation, i.e., the small particles are adhered to the bigger ones. This phenomenon could be proved by microscopic observations. As a consequence, the respirable dust concentration at the workplace also decreases (see Fig. 5.16). At the same time, the total dust concentration is not much changed.

There is fundamental difference in the streamline profile before a tube end depending on the flow direction. In the case of outflow, a turbulent free jet develops

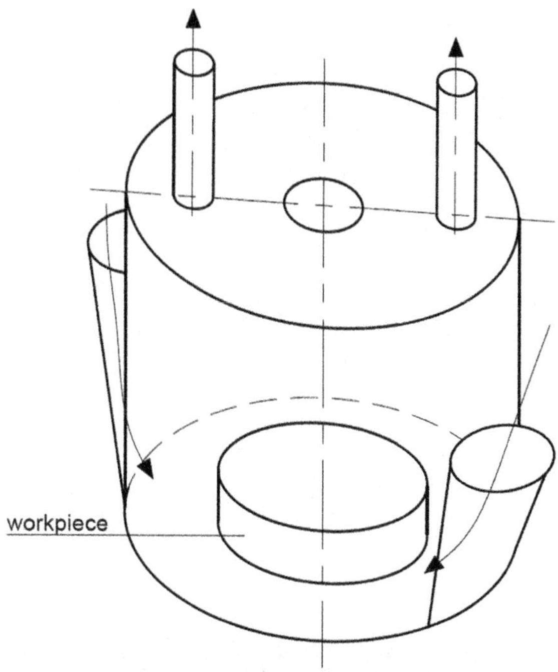

Fig. 3.5 Example of improved suction head of CNC machines

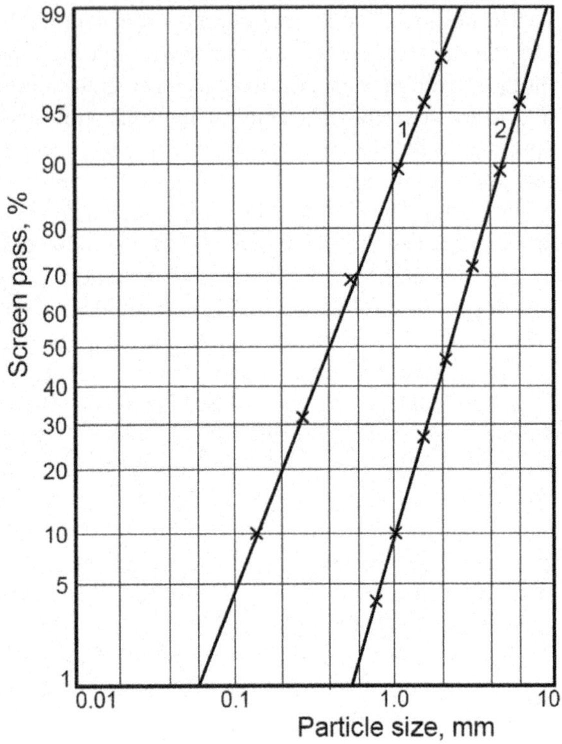

Fig. 3.6 Size distribution of oak dust extracted at a CNC machine for two different moisture contents. Milling tool of 60 mm dia., $e_z =$ 0.3 mm. 1. 10% m.c., 2. 18% m.c.

and its cross-section extends due to large-scale turbulent pulsations, Fig. 3.7a. In the case of suction, the airflow enters the tube from all directions and a quite different flow field develops, Fig. 3.7b. The main difference can be characterized by the action distance. The velocity of suction field in the tube axis at a distance of d (tube diameter) is only one-tenth of the tube velocity v_0. At blowing, one-tenth of the outlet velocity is obtained at a distance of $30 \cdot d$, Fig. 3.7c.

Practical importance of this phenomenon appears in suction hood design and in the outlet design of ventilating and the recirculated air. If the outlet diameter of recirculated air is large, the very efficient air blowing can cause draft effect. A slit opening with long and narrow cross-section has small equivalent diameter and herewith smaller action distance (Koanda aeration pipeline).

In the case of suction, the decrease of air velocity in the tube axis as a function of distance x can be calculated from the following equation

$$\frac{v_x}{v_0} = \frac{1}{1 + 7.7\left(\frac{x}{\sqrt{A}}\right)^{1.4}} = \frac{1}{1 + 9.12(x/d)^{1.4}} \tag{3.3}$$

where A is the tube cross-section. Please note that $x/\sqrt{A}$ is dimensionless and the equation is generalized as a function of x/d. The outflow velocity is simply given by

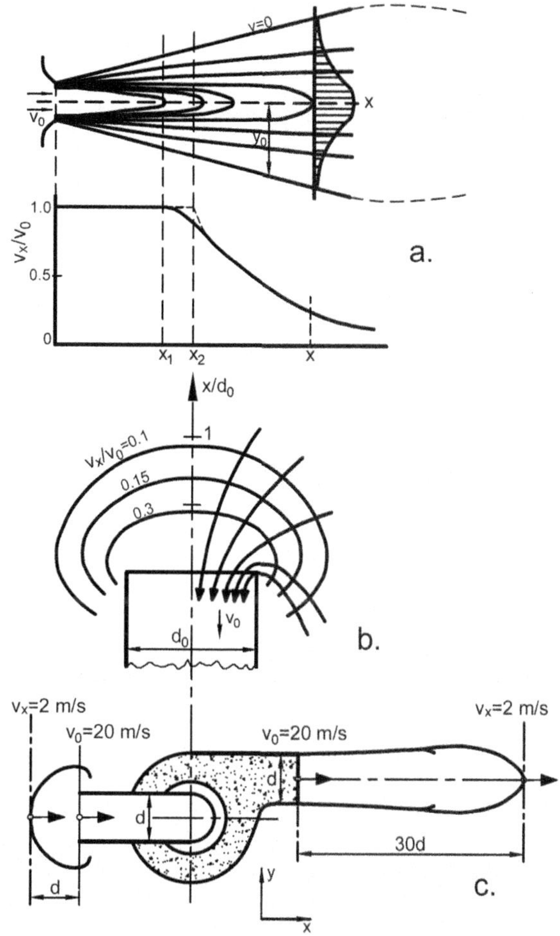

Fig. 3.7 Streamlines before a free tube end in the case of outflow (**a**) and inflow (**b**), and the action distance of suction and blowing (**c**)

$$v_0 = \sqrt{2g \cdot \frac{\Delta p}{\rho_0}} = 1.273 \cdot \sqrt{\Delta p}$$

where Δp means the over-pressure in the ventilation duct.

The generalization of theoretical and experimental results (Abramovich 1960) allows to calculate the turbulent free jets in simplified manner. The velocity decrease along the x-axis can well be described with the following dimensionless equation:

$$\frac{v_x}{v_0} = \frac{0.96 \cdot R}{a \cdot x}$$

with $a = 0.075$ for circular orifices of radius R. Taking $x = 30 \cdot d = 60 \cdot R$ distance, the velocity ratio is 0.21. Remembering the velocity distribution within the jet cross-section (see Fig. 3.7a), the expected average velocity is surely not more than $v_{xav}/v_0 = 0.1$. In ventilation systems, the use of circular inlet orifices is not favorable and the practice applies slit opening giving smaller equivalent diameters for a given cross-section. The developing flat-free jet has much more a two-dimensional behavior and the decrease of velocity along the x-axis is given by

$$\frac{v_x}{v_0} = \frac{1.2\sqrt{b_0}}{\sqrt{a \cdot x}} \tag{3.3a}$$

where b_0 is the half-width of the slit opening and the constant has the value of $a = 0.11$.

The equivalent diameter of the slit opening is calculated as

$$d_e = \frac{4A}{K}$$

where

A is the cross-section of the opening,
K is the circumference of the opening.

At a distance of $30 \cdot d_e$ the calculated velocity ratio is $v_x/v_0 = 0.33$. That means that in flat-free jets the rate of velocity decrease is slightly smaller compared to circular jets.

Conclusion: In order to capture most of the dust particles it is important:

- to utilize the particle motion given by the tool with proper placing of the suction pipe;
- to use a proper hood design, and in some cases tool design, to ensure sufficient air velocity around the tool. The sufficient air velocity is always higher than the terminal velocity of the particles.

In the practice individually adjusted suction hoods are used to improve the dust capture efficiency according to the above rules, Fig. 3.8.

Fig. 3.8 Individually adjusted extraction hoods (Leitz DFC®-Extraction hoods)

References

Г. Абрамович, *Теория турбулентных струй* (*Theory of Turbulent Free Jets*) (Физматгиз, Москва, 1960)
G. Sitkei, *Mechanics of Agricultural Materials* (Elsevier, Amsterdam, 1986), pp. 284–402
G. Sitkei, Particle motion with air resistance. Prog. Agric. Eng. Sci. 1–15 (2012)

Internet Websites

Gruing Aerotech System, https://www.cruing.com/aerotech/aerotech-system-e/
Leuco I-system, https://wtp.hoechsmann.com/cz/lexikon/21605/i-system
Leitz DFC®-Extraction Hoods, https://www.leitz.org/en-us/highlights/dfc-extraction-hoods/

Chapter 4
Pneumatic Conveying

Using various woodworking operations, the removed material in the form of dust and chips with a given particle size distribution is collected by air stream in a suction hood. It is convenient to use this air stream also for transporting the collected dust and chips in a pipeline to the storage bin. The aerodynamic resistance (drag) on the particles supplies the force for pneumatic transport: if this force exceeds the force required to displace the particles, the material starts to move with air. The driving force depends on the velocity difference between the air and particle therefore, the material motion velocity is always less than the air velocity.

A simplified layout of the pneumatic transport system is given in Fig. 4.1. The main components of the system are the suction hood, ventilator, straight tubes in horizontal and vertical positions, bent tubes, separation device, and storage bin.

4.1 State Diagram for Pneumatic Transport

In order to understand the mechanical behavior of granular material in an air stream, the general state diagram of pneumatic transport is often used.

It is convenient to study first the grain-air state diagram for transport vertically upwards in a tube. In the state diagram, the pressure drop per 1 m tube length is measured on the vertical axis, and the mean velocity of the airflow (relative to the cross-sectional area of the empty tube) is potted on the horizontal axis on a logarithmic scale, except for points in the immediate vicinity of the origin, Fig. 4.2 (Sitkei 1986).

The curve *E–G* represents the flow of air alone ($m_a = 0$) in the tube, and the curve *A–C* the aerodynamic resistance of the materials filling the tube completely, when in the stationary state. On charging the material to be transported the pressure drop increases, and therefore the curves take up positions above the curve *E–G*, depending on the concentration of material. When the air velocity decreases, the required pressure decreases for a while and then increases again. The reason is that

E. Magoss et al., *Dust Extraction and Handling in the Wood Industry*,
SpringerBriefs in Applied Sciences and Technology,
https://doi.org/10.1007/978-3-031-08915-2_4

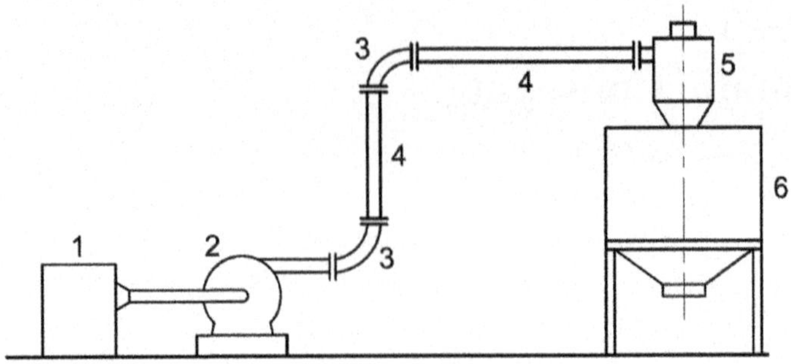

Fig. 4.1 The main components of a pneumatic transport system. 1—Woodworking machine, 2—ventilator, 3—bent tube, 4—tube, 5—cyclone, 6—storage bin

Fig. 4.2 State diagram for pneumatic transport

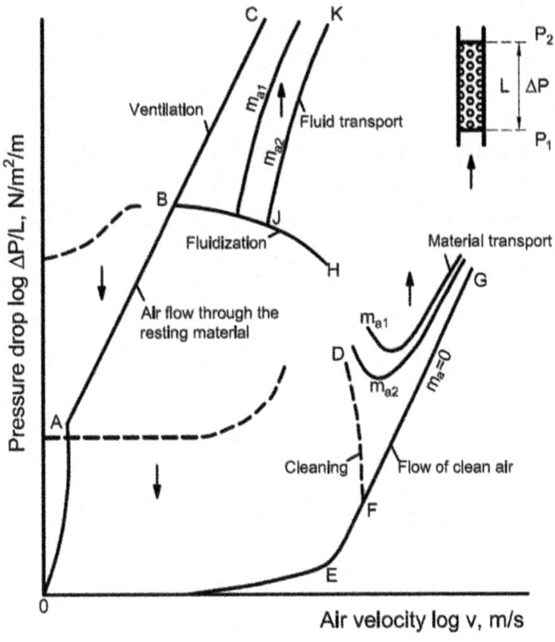

when the air velocity is decreased the uniformity of the flow is disturbed, bunching occurs in the tube, and the movement of both the air and the material transported becomes unsteady (Fig. 4.3). The line *F–D* represents equilibrium, with no transport. With further decrease in velocity, the bunching increases, and finally the tube is filled by the material. An equilibrium state may also be created here, without yet any transport. This state, which is indicated by the line *B–H* in the state diagram, is termed the fluidized state. Above the line *B–H* (e.g., on the line *J–K*), plug transport is possible. Left from the line *F–D* and under the line *B–H* no transport is possible, the material moves downwards (see arrow).

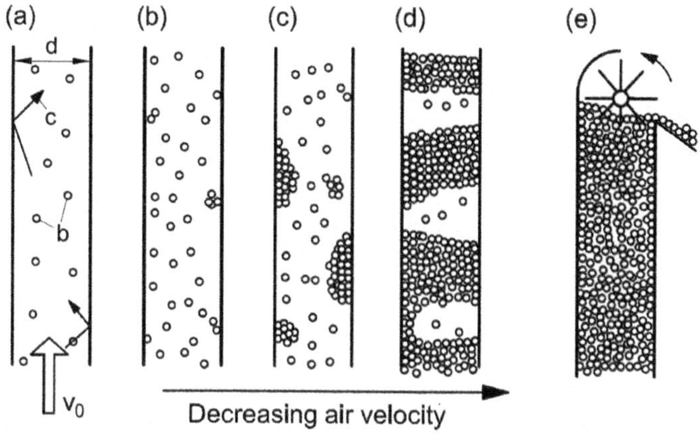

Fig. 4.3 Decrease in uniformity of transport with reduction of air velocity

The state diagram for horizontal transport is constructed similarly and the boundary curves *A–C* and *E–G* are identical. When the velocity decreases, similar phenomena appear in horizontal as in vertical transport: the development and appearance of the characteristic states are similar to those illustrated in Fig. 4.2.

4.2 Operational Characteristics

Concerning pneumatic transport of materials, first of all, the different velocities play the most important role:

- the average air velocity in the tube, v_0,
- particle motion velocity, v_m,
- the blow velocity on the particle giving the driving force, $w = v_0 - v_m$,

The lag of particles behind the air is characterized by the slip defined in the following manner:

$$s = \frac{v_0 - v_m}{v_0} \quad \text{or} \quad \frac{v_m}{v_0} = 1 - s \tag{4.1}$$

Due to the slip of particles, the material concentration in the tube depends on the charged material in the unit time (kg/s) and on the slip (Pápai 1954; Pattantyus 1954)

$$\mu_t = \frac{m_m}{m_0 \cdot (1 - s)} \tag{4.2}$$

where

m_m is the charged material, kg/s,

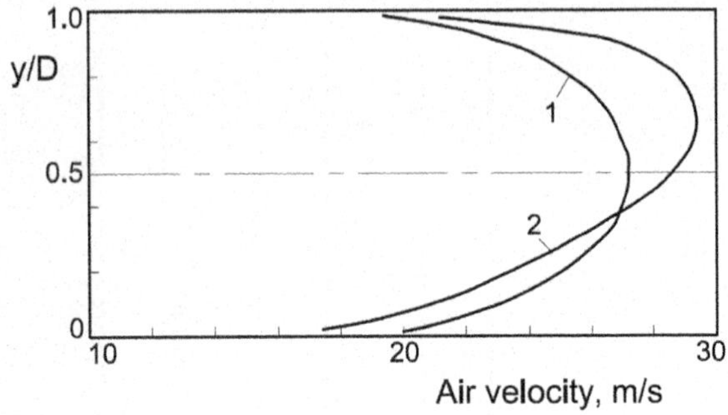

Fig. 4.4 Air velocity distribution in a tube without (1) and with material transportation (2)

m_0 is the air mass flow, kg/s.

The airflow is simply calculated as

$$m_0 = A_0 \cdot v_0 \cdot \rho_0 \tag{4.3}$$

where

A_0 is the tube cross-section, m^2,
ρ_0 is the air density, kg/m^3.

The material transported in a horizontal tube does not fill the space uniformly but is found mainly in the lower part of the cross-section, owing to gravity. The air velocity profile is greatly modified by the unequal material distribution as compared to the symmetrical velocity distribution prevailing at air transportation alone. Figure 4.4 shows measured examples for clean air and material transportation.

The concentration of transported material is distributed in the opposite sense and in the bottom part of the tube concentration may reach between three to five times the average concentration, Fig. 4.5 (Welschof 1962). The higher the transport throughput, the less the effect of air velocity on the distribution of material. From the above, it may be seen that in a horizontal transport tube the material moves mostly in the lower part, and the air, at a higher velocity, in the upper part of the tube. Consequently, the mean effective air velocity acting on the material is lower than the mean air velocity v_0 (Welschof 1962). An obvious consequence is that the minimum transport air velocity will increase with higher mass load m_m.

Another important operational characteristic is the minimum air velocity which ensures trouble-free operation without clogging. The minimum air velocity depends on several factors: particle properties such as size, shape, elasticity, and moisture content play the most important role, but the tube diameter and transport concentration have also definite influences (Pápai 1954; Segler 1951).

The elastic properties of particles fundamentally determine the nature of transport losses: a more elastic particle (crops as wheat or maize) mainly causes impact losses while soft particles with minimum rebound (such as wood particles) cause friction losses.

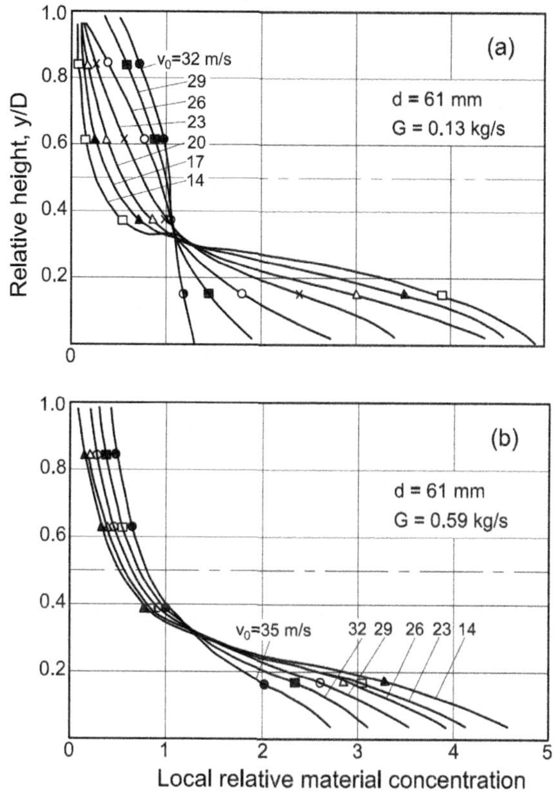

Fig. 4.5 Relative local material concentration as a function of height in transport tube during transport of wheat

The moisture content is a major influencing factor almost in all cases. Even crop products with high moisture content become less elastic with increased damping. Furthermore, the moisture content increases the friction coefficient (see in Sect. 1.4) and hereby the friction losses. Therefore, a moist material transported is more inclined to non-uniform distribution in the tube cross-section requiring higher minimum air velocity for trouble-free operation. Figure 4.6 shows the high difference in the minimum air velocity for maize transportation with different moisture contents (replotted from Flatow and Siegel 1969).

The minimum air velocity for air-dry dust and chips is given in Fig. 4.7 as a function of transport concentration. The curve follows the empirical equation

$$v_{\min} = 16.5 + 3.14 \cdot \mu_t^{0.6} \tag{4.4}$$

which is valid in the tube diameter range of 100 and 160 mm. For higher tube diameters it is recommended to increase the minimum air velocity slightly (Segler 1951). In the figure, the material velocity is also given (calculation of slip is described later in Sect. 4.4.1).

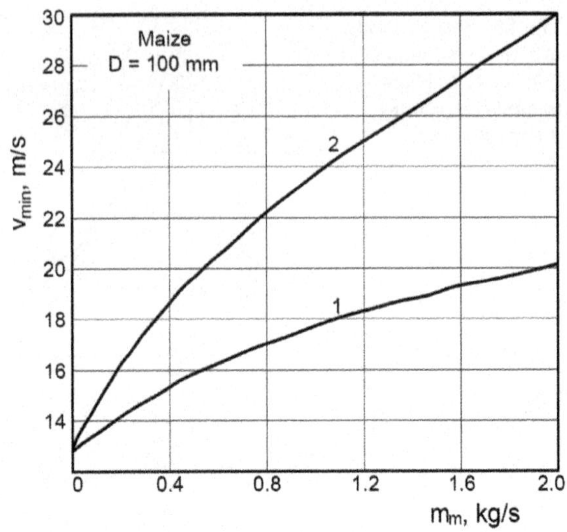

Fig. 4.6 Minimum air velocity for maize transportation as a function of mass load for air-dry (1) and moist maize with 37% moisture content (2)

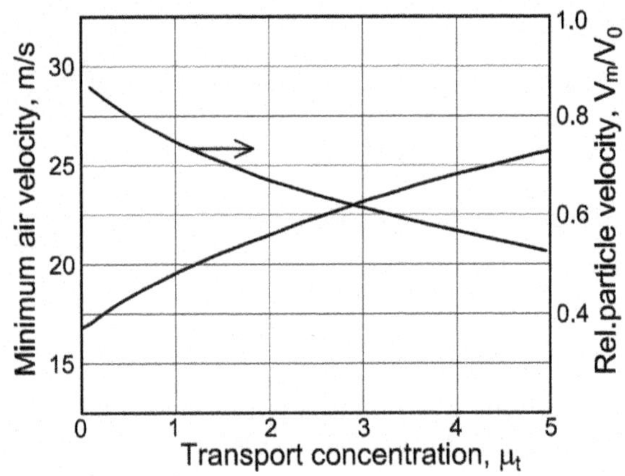

Fig. 4.7 Minimum air velocity for air-dried dust and chips as a function of transport concentration and the corresponding material velocity

4.3 Resistance to Airflow

In a real flow, independently of laminar or turbulent, shear stresses are developed due to internal friction and velocity gradient across the tube. The velocity profile along a given radius depends on the nature of flow: in laminar flow, a second-order parabola holds, while in turbulent flow a higher-order parabola, Fig. 4.8. The different forms of flow were discovered in 1883 by Reynolds. At low velocities local elemental disturbances can be balanced by molecular forces and, therefore, the individual streamlines run in an orderly manner side by side. That is the laminar flow. At higher velocities, the disturbances cannot be balanced by molecular forces and nature created an additional motion of small volumes to balance disturbances. That is the turbulent flow

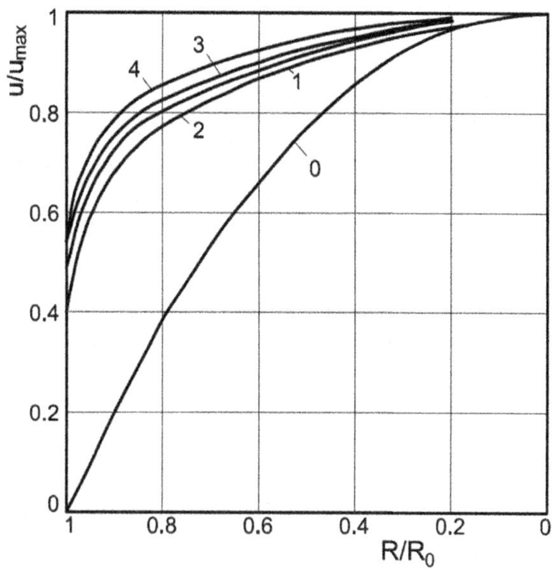

Fig. 4.8 Velocity distribution along the tube radius, 0—laminar flow, 1—$Re = 4 \times 10^3$, 2—$Re = 4 \times 10^4$, 3—$Re = 4 \times 10^5$, 4—$Re = 3.2 \times 10^6$

and the individual fluid volumes execute fluctuating motion in all directions around some mean flow path, for us, in an irregular way. On the tube wall, the velocity is always zero due to molecular forces.

The nature of flow is more exactly characterized by the dimensionless Reynolds number defined by the mean velocity v_0, the tube diameter D, and the kinematic viscosity of fluid:

$$Re = \frac{v_0 \cdot D}{\nu}$$

The critical Reynolds number separating the two forms of flow is 2300. In practical cases, we always encounter turbulent flow.

In order to calculate friction losses and the pressure drop along the tube length, instead of the shearing stresses, the friction coefficient λ is used in the following form:

$$\Delta p = \lambda \cdot \frac{L}{D} \cdot \frac{\rho_0}{2} \cdot v_0^2 \qquad (4.5)$$

where

D is the tube diameter,
ρ_0 is the air density at a given temperature.

Early detailed experiments have shown that the surface roughness definitely influences the friction coefficient (Nikuradse 1932, 1933). As a characteristic number, the ratio of the average height of the roughness elements k_r to the tube radius R is used.

Figure 4.9 shows the experimental results for different roughness ratios R/k_r, as a function of Reynolds number. For sufficiently large *Re*-numbers, the curves become horizontal, indicating that the pressure drop increases in proportion to the square of the air velocity.

For quick estimate of the λ/D ratio an auxiliary diagram is worked out, Fig. 4.10 (Tauber 1978). This diagram supplies friction values approximately corresponding

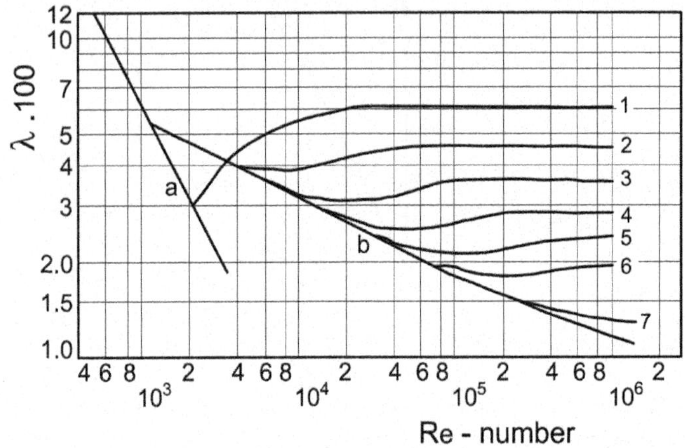

Fig. 4.9 Friction coefficients for rough tube walls. **a** Laminar flow, **b** turbulent flow in hydraulically smooth tubes. 1—$R/k_r = 15$, 2—$R/k_r = 30$, 3—$R/k_r = 80$, 4—$R/k_r = 125$, 5—$R/k_r = 250$, 6—$R/k_r = 500$, 7—$R/k_r = 1300$

Fig. 4.10 The λ/D ratio for quick estimate as a function of tube diameter for different air velocities

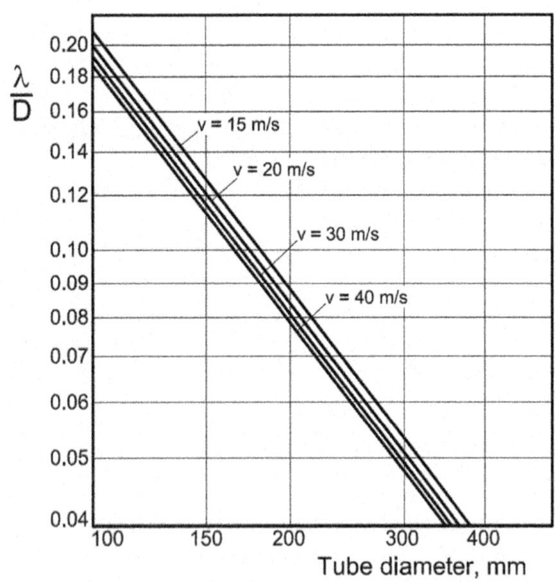

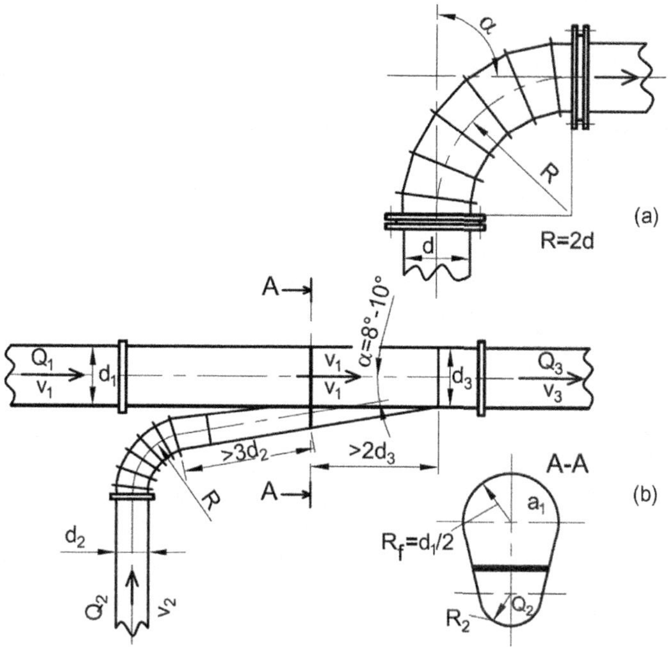

Fig. 4.11 Structural layout of bent tube (**a**) and a joining element connecting a suction tube to the main transport tube (**b**)

to curve No. 6 in Fig. 4.9. For smooth tubes somewhat lower friction values may be taken into account.

Further elements of the system are the bent tubes and joining elements connecting suction tubes to the main transport tube, Fig. 4.11. These elements cause local losses, which are calculated, using loss coefficients ξ_i, with the following equation:

$$\Delta p_i = \xi_i \cdot \frac{\rho_0}{2} \cdot v_0^2 \tag{4.6}$$

Bent tubes generally have an angle of 90° between inlet and outlet, and in this case, the loss coefficient has a value of 0.15. For smaller angles, this value decreases, more or less, in proportion to the decrease of angle.

For joining elements, there is an important condition that both flow velocity in the meeting point would be the same. Using small joining angle of 8 to 10°, the local loss may be disregarded.

4.4 Resistance of Material Transportation

In order to transport material in an air stream, an additional driving force (pressure drop) is needed. The energy of pressure drop depends on many factors: shape, size, moisture content, and elastic–plastic behavior of particles are the most important properties. We distinguish here two different particle behavior:

- Particles having rounded and smooth surfaces with considerable elastic behavior, such as crop products. During their motion in a tube, they impact with the tube wall and each other. The particles encounter impact losses due to damping.
- Particles with irregular shape, hairy surface, and nearly plastic behavior. Due to the very low rebound, the particles move mostly with sliding on the tube wall causing friction losses.

Although the two types of losses may appear together, their separation can only be made with unreliable assumptions. Furthermore, the slip behavior of transported materials for the two kinds of losses differs fundamentally and the determination of a common slip seems to be difficult. The slip is, however, an important operational characteristics determining the transport concentration. Therefore, we will treat the above losses as pure impact and friction ones.

4.4.1 Resistance Due to Impact

Crop products such as wheat and maize more or less correspond to the impact behavior with some damping. In his classical work, Pápai has conducted detailed experimental and theoretical analysis of impact losses using air-dry wheat (Pápai 1954, 1966). The equilibrium equation of motion can be given in the following form:

$$c_e \cdot f_0 \cdot \frac{\rho_0}{2} \cdot w^2 = k_i \cdot \frac{m_1 \cdot v_m}{2} \quad \text{and} \quad w + v_m = v_0 \tag{4.7}$$

where

c_e is the drag coefficient,
f_0 is the cross-section of particle,
ρ_0 is the density of the air,
m_1 is the particle mass,
w is the blow velocity of the air on the particle,
v_m is the particle velocity,
v_0 is the air velocity,
k_i is the impact coefficient.

The mass of the particle can be expressed, using an equivalent particle diameter d, in the following manner:

$$m_1 = \frac{d^3 \cdot \pi}{6} \cdot \rho_m$$

where ρ_m means the density of the particle. Equation (4.7) can be transformed to the following form

$$\frac{w}{v_m} = \frac{s}{1-s} = \sqrt{\frac{2 \cdot k_i \cdot \rho_m \cdot d}{3 \cdot \rho_0 \cdot c_e}} \tag{4.8}$$

The dimensionless number under the square root uniquely determines the slip of the particles. Furthermore, if we remember the equation of terminal velocity of a particle, Eq. (4.8) can be modified as

$$\frac{s}{1-s} = \sqrt{\frac{k_i}{2g}} \cdot v_{cr} \tag{4.8a}$$

where v_{cr} means the terminal velocity. Equation (4.8) is demonstrated in Fig. 4.12. As an example, for wheat grain the following values are valid: $k_i = 0.085$ 1/m, $\rho_m = 1380$ kg/m^3, $d = 3.5$ mm, $c_e = 0.5$, and with these values $v_{cr} = 10.054$ m/s and $s = 0.4$ in agreement with experimental results. It is worth to mention that the slip is influenced by the impact coefficient k_i. Latter depends on the moisture content which increases the damping at impact. As a consequence, moist crops can be transported with higher slips.

The pressure drop of the material transport due to impact losses is calculated (Pápai 1954) as

$$\frac{\Delta p_m}{L} = \frac{k_i \cdot (1-s)}{2} \cdot \frac{m_m \cdot v_0}{A_0} \quad \frac{\text{N/m}^2}{\text{m}} \tag{4.9}$$

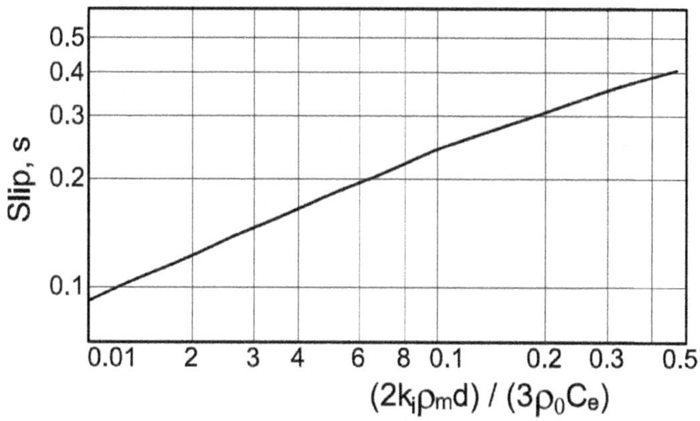

Fig. 4.12 Slip of the pneumatic transport as a function of dimensionless number. Impact losses

where A_0 is the tube cross-section.

In order to calculate the total energy consumption, the air resistance according to Eq. (4.5) should be added. The efficiency of material transportation may be checked by the following equation:

$$\eta = \frac{V \cdot (\Delta p_m + \Delta p_0)}{m_m} \quad \text{N m/kg} \tag{4.10}$$

where V means the transported air volume in the unit time, m^3/s.

It is worth to mention that for dilute phase transport the practice also today uses the simple Gasterstädt equation in the following form

$$\Delta p_{t0t} = \Delta p_0 + \Delta p_m = \Delta p_0 \cdot (1 + k_0 \cdot \mu_t) \quad \text{and} \quad k_0 = \frac{\Delta p_m}{\Delta p_0 \cdot \mu_t}$$

where the loss factor k_0 takes the additional pressure drop into account. Using Eq. (4.9), k_0 can be expressed by system parameters as follows

$$k_0 = \frac{k_i \cdot (1 - s)^2 \cdot D}{\lambda}$$

Example. Wheat is transported in a tube $D = 130$ mm with $k_i = 0.085$ 1/m, $\lambda/D = 0.145$ and $s = 0.4$. For this case we obtain $k_0 = 0.211$.

The friction coefficient of tube λ is more or less constant and, therefore, the tube diameter D and the impact loss factor k_i are the important influencing parameters. The impact loss factor is mainly influenced by the moisture content.

In the earlier time, $k_0 = 0.3$ was generally proposed for crop products independently of system parameters. Now it is possible to calculate k_0 values more accurately.

4.4.2 Resistance Due to Friction

Particles originated in the wood industry are produced by wood machining, either with knife or abrasive belt (sanding) and sometimes with hammer mill. The chip formation is associated with brittle fracture resulting in irregular shape, hairy and rough surface. These particles have almost plastic behavior with minimum rebound after impact with the tube wall and each other. As a consequence, the material is mostly sliding in the lower part of a horizontal tube and its resistance is determined by friction forces.

The driving force is also here the aerodynamic drag. It is probably, however, that, apart from very low transportation concentrations, not single particle but coagulated particles forming bigger groups have an irregular face perpendicular to the air velocity and, therefore, the drag coefficient may have values around one. The particle groups have no stable shape during their motion, which may cause unsteady motion with inclination to local clogging. It may be expected that the required minimum air velocity is higher compared to the impact resistance.

The aerodynamic force acting either on a single particle or a group of particles can be given as

$$F_a = c_e \cdot f_0 \cdot \frac{\rho_0}{2} \cdot w^2 \quad \mathrm{N}$$

This force is in equilibrium with the friction forces acting along 1 m tube length. Using the expression of pressure drop (Pápai 1954).

$$\Delta p_m = \frac{k_s \cdot m_m \cdot g \cdot L}{A_0 \cdot v_m} \quad \mathrm{N/m^2} \tag{4.11}$$

where k_s is the friction coefficient for the transported material.

The corresponding friction force for $L = 1$ m is given by the equation

$$F_s = \frac{k_s \cdot m_m \cdot g \cdot f_0}{A_0 \cdot v_m}$$

Taking $F_a = F_s$ for equilibrium, the above equations yield:

$$\frac{k_s}{c_e} = \frac{\rho_0 \cdot A_0 \cdot w^2 \cdot v_m}{2 \cdot m_m \cdot g} = \frac{\rho_0 \cdot A_0}{2 \cdot g \cdot m_m} s^2 \cdot (1 - s) \cdot v_0^3 \tag{4.12}$$

or with further transformation

$$\frac{k_s}{c_e} = \frac{(s \cdot v_0)^2}{2g \cdot \mu_t} \tag{4.12a}$$

Concerning the very rough surface of particles and their motion in bigger groups and bundles, the drag coefficient c_e is near to one and, therefore, in the following we always use $c_e = 1.0$. The friction coefficient k_s, takes also the possible deviation in c_e into account. Seemingly, Eq. (4.12a) is dimensionally incorrect, because the unit length $L = 1.0$ m is omitted. In order to avoid dimensional errors, in this case, k_s takes over the dimension of length.

In the above equation we have two independent variables s and v_0, and, therefore, it can only be used with some combination of s and v_0. An evaluation of Eq. (4.12a) is given in Fig. 4.13. The constant slip-lines clearly show that with increasing transport concentration the air velocity should be increased in order to ensure acceptable slip values and stable, clogging-free operation.

In order to obtain a second independent equation to Eq. (4.12a), the theorem of minimum potential energy may be used. In this case, we seek the energy minimum which is required to transport 1 kg material for a distance $L = 1$ m. The efficiency of transportation (N m/kg m) is given by the equation (see Eq. 4.10):

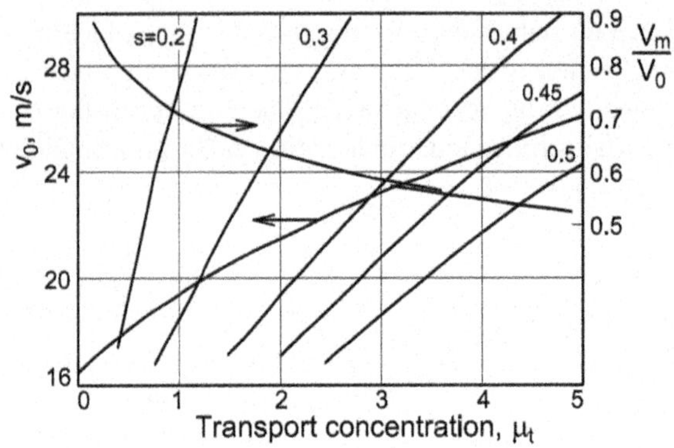

Fig. 4.13 Constant slip-lines in the v_0–μ_t coordinate system and variation of relative material velocity v_m/v_0 as a function of transport concentration for a given minimum air velocity, $k_s = 1.5$

$$\eta = \frac{V \cdot (\Delta p_m + \Delta p_0)}{m_m}$$

where V is the air volume transported in the unit time, $V = A_0 \cdot v_0$.

The pressure drops are given by Eqs. (4.5) and (4.11). Note that the pressure drop for material transportation can be expressed with the transport concentration as follows

$$\frac{\Delta p_m}{L} = k_s \cdot \gamma_0 \cdot \mu_t \qquad (4.13)$$

where γ_0 is the specific weight of the air, N/m^3.

Substitution into Eq. (4.10) yields

$$\eta = \frac{v_0}{v_0 - \sqrt{k_s \cdot 2g \cdot \mu_t}} \cdot \left[k_s \cdot g + \frac{(\lambda/D) \cdot v_0^2}{2 \cdot \mu_t} \right] \qquad (4.10a)$$

Taking the derivative of Eq. (4.10a) in respect to μ_t and equal to zero

$$\frac{\partial \eta}{\partial \mu_t} = 0 \qquad (4.14)$$

It is advisable to seek the minimum values numerically in the following way: for a chosen value of μ_t we seek that air velocity v_0 which gives the minimum energy of transportation. Evaluation of Eq. (4.10a) is plotted in Fig. 4.14 for a tube diameter of 130 mm (Magoss et al. 2021).

The optimum air velocity increases steeply as a function of concentration. The corresponding slip is 0.45 (45%) and the relative material velocity is 0.55 independently of k_s value. Sadly, the optimum air velocities for lower concentrations are small and they do not fall into the feasible range of clogging-free operation. In practice, the required minimum air velocity (v_{min}) varies in a parabolic fashion depending, first of all, on the moisture content. The corresponding slip increases the v_m/v_0 ratio and

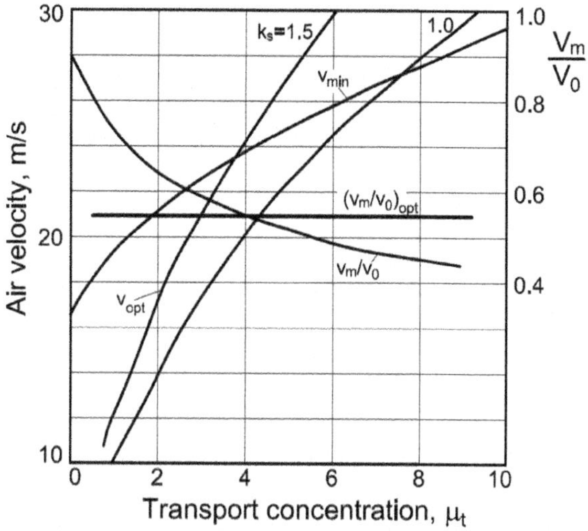

Fig. 4.14 Optimum air velocity for minimum energy consumption for $k_s = 1.0$ and 1.5, minimum air velocity for clogging-free operation, and the corresponding v_m/v_0 ratio as a function of transport concentration

decreases with higher transport concentrations. The optimum energy consumption, for $k_s = 1.5$, amounts to 45.9 N m/kg m which, using optimum air velocities, would be independent of concentration. In practice, this optimum energy consumption is achievable only at a concentration where $v_{\min} = v_{\text{opt}}$, in our case for $\mu_t = 4$.

In the possession of coherent v_0 and μ_t values, the slip and material velocity may be calculated as

$$v_m = v_0 - \sqrt{k_s \cdot 2g \cdot \mu_t} \quad \text{and}$$

$$s = \frac{\sqrt{k_s \cdot 2g \cdot \mu_t}}{v_0} = \frac{w}{v_0} \tag{4.15}$$

or

$$\frac{v_m}{v_0} = 1 - s = 1 - \frac{\sqrt{k_s \cdot 2g \cdot \mu_t}}{v_0}$$

In practice, the input data are the charged mass m_m and the air volume V_0. m_m may be calculated from machining operational parameters while V_0 is generally prescribed by the machine manufacturer in order to capture the dust particles in the suction hood in a sufficient extent. Using Eq. (4.2), the relationship between μ_t and m_m is given as

$$\mu_t = \frac{m_m \cdot v_0}{\rho_0 \cdot V_0 \cdot v_m} = \frac{m_m \cdot v_0}{\rho_0 \cdot V_0 \cdot \left(v_0 - \sqrt{k_s \cdot 2g \cdot \mu_t}\right)} \tag{4.16}$$

The minimum air velocity for clogging-free operation is the function of transport concentration μ_t (see Eq. 4.4). Equation (4.16) is implicit for μ_t and for a quick estimate, the following simple approximation may be used

$$\mu_t = \frac{1.416}{\rho_0 \cdot V_0} \cdot m_m^{1.06} \tag{4.16a}$$

Example: $m_m = 0.035$ kg/s, $V_0 = 1200$ m³/h $= 0.333$ m³/s, from Eq. (4.16) $\mu_t = 0.096$ and $v_0 = 17.3$ m/s. Equation (4.16a) supplies $\mu_t = 0.1$. The slip is obtained from Eq. (4.15) and $s = 0.096 = 9.6\%$. The required tube diameter is 0.157 m, the pressure drops are $1.74 + 21.95 = 23.7$ N/m^2 per meter.

Second Variant

$m_m = 0.035$ kg/s, $V_0 = 800$ m^3/h $= 0.222$ m^3/s, $\mu_t = 0.143$, $v_0 = 17.48$ m/s, $d = 0.127$ m, $\lambda/D = 0.15$, $\Delta p_0 = 28.65$ N/m^2, $\Delta p_m = 2.68$ N/m^2, $\Delta p_z = 31.33$ N/m^2. Smaller tube diameter but higher pressure drop is needed.

In the above discussion, we have generally used a friction coefficient $k_s = 1.5$ which gives acceptable results for air-dry dust and chips. For moist saw chips with 37–40% moisture content k_s values up to 2.5 have been measured (Boronkai 1993).

There is a minimum air velocity at which no material transportation occurs. Obviously, in this case, we have $v_m = 0$ and the slip $s = 1$, and Eq. (4.15) yields the following equality:

$$v_{0\,\min} = \sqrt{k_s 2g \cdot \mu_t} \tag{4.15a}$$

which is graphically demonstrated in Fig. 4.15. For common concentrations ($\mu_t \leq 1$) the minimum air velocity varies well below 10 m/s, but for higher concentrations, it

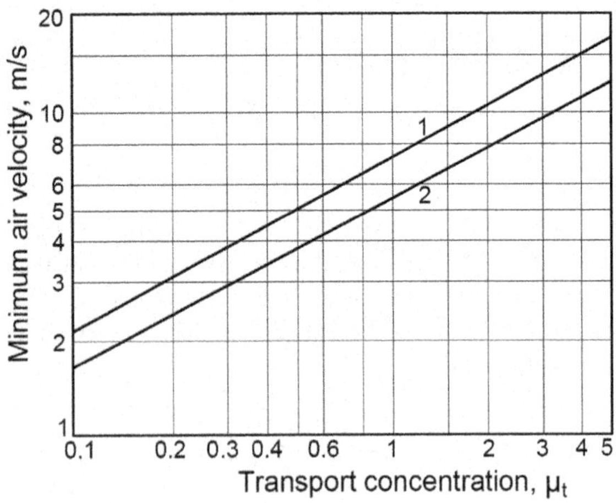

Fig. 4.15 Minimum air velocity with no material transportation as a function of material concentration. 1—$k_s = 2.5$, 2—$k_s = 1.5$

may have values around 15 m/s. This theoretical finding supports the practical experience that the safe transport of air velocity increases with the material concentration and dust moisture content. It is interesting to note that the empirical Eq. (4.4) based on practical observations has almost the same exponent as a function of transport concentration μ_t. The higher velocity values of Eq. (4.4) correspond to a chosen coefficient of safety which is always used in the practice.

In the case of low transport concentrations, the simple equation of Gasterstädt (1924) is often used. This approach calculates the total pressure drop using a multiplier k_0 in the following form

$$\Delta p_{t0t} = \Delta p_0 \cdot (1 + k_0 \cdot \mu_t)$$

where Δp_0 means the pressure drop for clean air. Keeping in mind Eq. (4.13), k_0 can be related to our more detailed calculation method as follows

$$k_0 = \frac{k_s \cdot \gamma_0}{\Delta p_0} = \frac{2 \cdot g \cdot k_s \cdot D}{\lambda \cdot v^2}$$

or with known tube diameter

$$k_0 = \text{constant} \cdot \frac{k_s}{v^2}$$

where v is the air velocity. Evaluation of the above equation for $d = 130$ mm tube diameter is given in Fig. 4.16.

Some experimentally determined k_0 values are known (Tauber 1978) and they fall into range between 0.6 and 0.8. Figure 4.16 clearly shows the strong dependence on the air velocity and a definite dependence on the equivalent friction coefficient k_s. The experimental results mentioned before fall into the feasible range in Fig. 4.16. The derived equation and the figure facilitate to select reliable values for k_0 with known input data.

4.4.3 Other Resistance Components of the System

In a real transport system, there are further resistance components due to

- acceleration of the captured particles to the stationary material velocity v_m,
- lifting the material in a vertical tube with the stationary material velocity v_m,
- material deceleration in bent tubes and its subsequent acceleration to the stationary transport velocity v_m,
- considerable local resistance in suction hoods, depending on their construction.

The pressure drop required for acceleration is calculated as

Fig. 4.16 Variation of coefficient k_0 as a function of air velocity and resistance coefficient k_s. Tube diameter is 130 mm

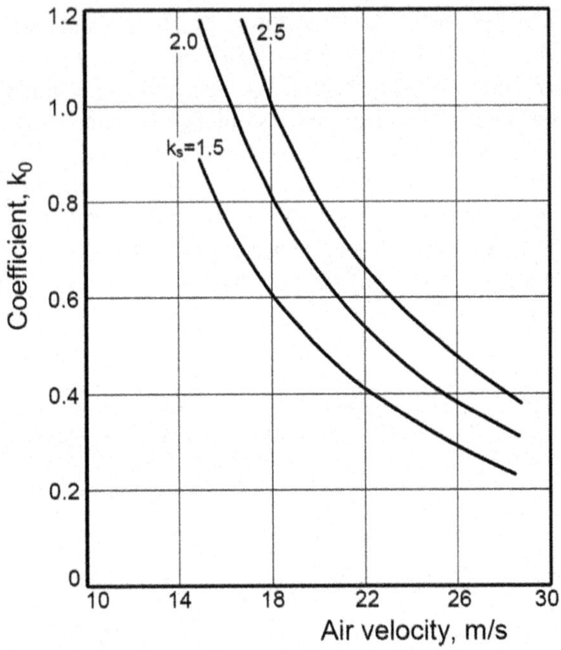

$$\Delta p_m = k_m \cdot \frac{m_m \cdot v_m}{A_0} \tag{4.17}$$

where the loss factor can be taken with $k_m = 1.0$. Using v_m in Eq. (4.17), we suppose that the initial particle velocity is zero. In many cases, however, the tool edge is acting as a throwing device giving an initial particle velocity in the direction of tube axis. In this case, the initial velocity decreases the required acceleration.

In vertical tube, the material should be lifted against gravity. The additional pressure drop is given by the equation

$$\frac{\Delta p_g}{L} = \frac{m_m \cdot g}{A_0 \cdot v_m} \tag{4.18}$$

Practical observation is that this loss in transportation of dust and chips not fully appears. An explanation is for it that in vertical tubes the distribution of particles is more uniform and the friction losses decrease compared to the horizontal tube. Therefore, for smaller vertical lengths, this loss may be neglected (Tauber 1978).

In bent tubes, located horizontally or vertically, the particles are sliding mostly on the tube wall suffering friction losses. Consequently, the outlet velocity of particles is roughly the half compared to the stationary inlet velocity v_m. The particles must again be accelerated to their stationary velocity. The acceleration takes place after the bent tube. The additional pressure drop is calculated with Eq. (4.17) using $v_m/2$ as an approximation (Kovács 1964):

$$\Delta p_b = \frac{m_m \cdot v_m}{2A_0} \tag{4.19}$$

Suction hoods in their construction forms are adjusted to the given woodworking machine and their main function is to envelop the tool and develop a flow field around the tool. This airflow is carrying the dust particles into the suction tube.

The local resistance of hoods is generally high and it is calculated according to Eq. (4.6). Their resistance coefficient is always experimentally determined and generally given by the manufacturer. Values between $\xi_i = 2$ and 3 are common.

4.5 Energy Consumption and Efficiency

The energy consumption and efficiency of a given transport system, similarly to Eq. (4.10), can be analyzed with the following general equation

$$\eta = \frac{V_0 \cdot \Sigma \Delta p_i}{m_m} \tag{4.20}$$

where all pressure drop components of the system are summed and related to the mass of transported material. This gives a real picture about a given built system but it is less suitable for comparison due to the differences among systems. Therefore, for comparison, it is more favorable to use Eq. (4.10) containing the air and material transport in a straight horizontal tube related to the unit length ($L = 1$ m). Concerning first impact losses, the pressure drop equation for material transport has the form:

$$\frac{\Delta p_m}{L} = k_p \cdot \frac{m_m \cdot v_0}{A_0} = k_p \cdot \mu_t \cdot \rho_0 \cdot v_0^2 \cdot (1 - s) \tag{4.21}$$

with

$$k_p = k_i \cdot \frac{1 - s}{2}$$

while the pressure drop due to airflow is given by Eq. (4.5).

Using Eqs. (4.5) and (4.9), the efficiency of transportation can be written in the following final form

$$\eta = v_0^2 \cdot \left[k_p + \frac{\lambda / D}{2\mu_t \cdot (1 - s)} \right] \tag{4.22}$$

For the evaluation of the above equation, we use the following input data: the minimum air velocity as a function of transport concentration is given by the equation.

$$v_0 = A + B\mu_t = 15 + 0.87\mu_t \text{ m/s}.$$

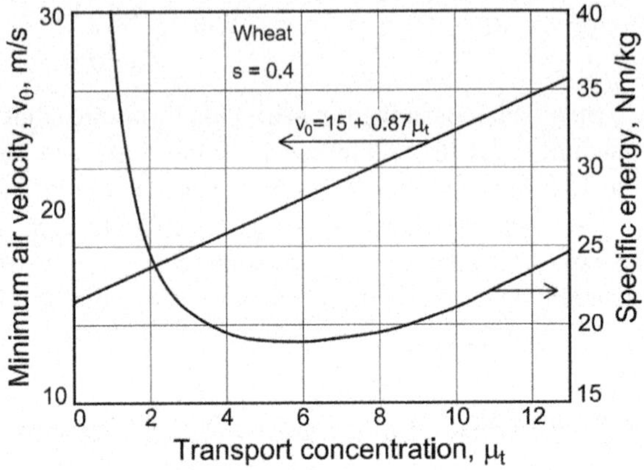

Fig. 4.17 Minimum air velocity and the specific energy as a function of transport concentration for air-dry wheat

In order to determine the optimum transport concentration, we substitute the expression of v_0 into Eq. (4.22), take its derivative in respect to μ_t and equate to zero, the following third-order equation is obtained:

$$2k_p \cdot B^2 \cdot \mu_t^3 + \left(2k_p \cdot A \cdot B + \frac{B^2 \cdot (\lambda/D)}{2(1-s)}\right) \cdot \mu_t^2 - \frac{A^2 \cdot (\lambda/D)}{2(1-s)} = 0 \quad (4.23)$$

Using the following input data: tube diameter $D = 130$ mm, $\lambda/D = 0.145$ (see Fig. 4.10), $k_i = 0.085$ 1/m and $k_p = 0.0255$ 1/m, the following equation is to be solved:

$$0.0386 \cdot \mu_t^3 + 0.757 \cdot \mu_t^2 - 27.187 = 0$$

which supplies $\mu_{topt} = 5.3156$ in accordance with the detailed calculations.

Evaluation of Eq. (4.22) is shown in Fig. 4.17, which shows a minimum between concentration values of 5–6. The exact value is given by the solution of Eq. (4.23). The curve may be regarded as flat in wide range of concentrations. For quite small concentrations, however, the specific energy consumption increases steeply.

Note that Eq. (4.22) has no minimum if v_0 is constant but decreases continuously with increasing concentration.

Concerning friction losses, the pressure drop for material transportation is given by Eq. (4.13) and the energy efficiency by Eq. (4.10a). Here we use the minimum air velocity equation as Eq. (4.4)

$$v_0 = 16.5 + 3.14 \cdot \mu_t^{0.6}$$

and the friction loss coefficient for air-dry chips with $k_s = 1.5$. The results of evaluation are demonstrated in Fig. 4.18 using numerical method.

First, it is striking that the specific energy consumptions are roughly doubled compared to impact losses. The curve is also here flat in a wide range of concentrations

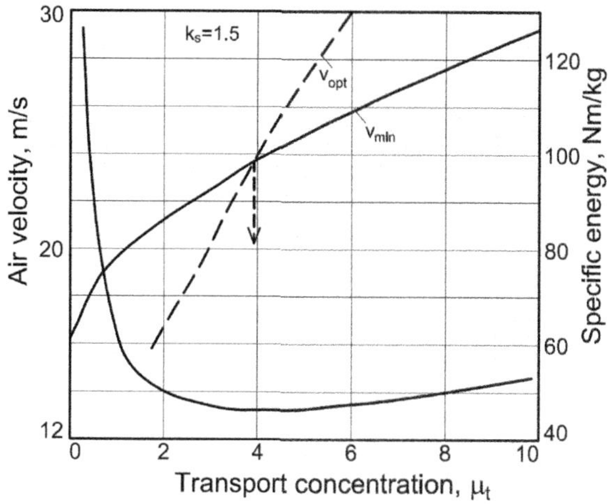

Fig. 4.18 Specific energy consumption for friction losses as a function of transport concentration for a given v_{min} function. The minimum is 45.9 N m/kg m at $\mu_t = 3.88$

but for small concentrations, it increases steeply. The minimum energy consumption is obtained at a point where the air velocity equals the theoretical optimum velocity (23.6 m/s), see in Fig. 4.14.

The quantity of produced chips by a woodworking machine is determined by the width and depth of cut and feed speed and the wood density. The expected maximum chip quantity does not exceed the 0.1 kg/s value but it is generally much less. Whereas the required air volume is rarely less than 800 $m^3/h = 0.222\ m^3/s$. As a consequence, the transport concentration is in the range of 0.1–0.3. In this case, the aim of the pneumatic system is much more the effective dust collection preventing air pollution at the working place. At the same time, the length of transportation should be as short as possible.

References

L. Boronkai, A faiparban alkalmazott por-forgács elszívó és szállítórendszerek méretezésének néhány kérdése (Pneumatic conveying systems in the wood industry). Ph.D. dissertation, Sopron (1993), pp. 39–85

J. Flatow, W. Siegel, Pneumatische Förderung von Körnermais in waagerechten Rohren. Grundlagen der Landtechnik, Bd. 19, Nr. 4, S. 125–128 (1969)

L. Gasterstädt, Die experimentelle Untersuchung des pneumatishen Förderganges. VDI Forschungsarbeiten, No. 265 (1924)

L. Kovács, Berechnung des Druckabfalls in 90° horizontal Krümmern pneumatischer Getreideförderleitungen. Periodica Polytech. Mech. Eng. (4), 447–467 (1964)

E. Magoss, G. Sitkei, Z. Kocsis, Collecting and handling of dust and chips in the wood industry. Progress Report No. 9, Published in 2021 by the Department of Wood Engineering, University of Sopron, Hungary (2021), p. 66

J. Nikuradse, Gesetzmässigkeiten der turbulenten Strömung in glatten Rohren. VDI Forsch. Heft, Nr. 356 (1932)

J. Nikuradse, Strömungsgesetze in rauhen Rohren. VDI Forsch. Heft, Nr. 361 (1933)

L. Pápai, *Pneumatikus gabonaszállítás* (*Pneumatic Grain Conveying*), vol. 12 (Magyar Tudományos Akadémia VI. Oszt. Közlemény, 1954), pp. 319–363

L. Pápai, Geschwindigkeits- und Druckverhältnisse bei waagerechter pneumatischer Förderung. Periodica Polytech. TU Budapest 397–415 (1966)

Á.G. Pattantyus, Pneumatic conveying. Acta Technica Hung. (1–2), 129–177 (1954)

G. Segler, Pneumatic grain conveying with special reference to agricultural application. Published by Prof. Dr. Ing. G. Segler, Braunschweig, Germany, S. 26–68 (1951)

G. Sitkei, *Mechanics of Agricultural Materials* (Elsevier, Amsterdam, 1986), pp. 284–402

Б. Таубер, *Внутризаводский транспорт* (*Internal Transport in Wood Industry*) (Машгиз, 1978)

G. Welschof, Pneumatische Förderung bei grossen Fördergut.koncentrationen. VDI Fohrschungsheft 492 (1962)

Chapter 5
Separation of Particles from the Airflow

Efficiency of dust separation is a major factor in environmental pollution and ventilation requirement of workrooms if air recirculation is used. New approaches are introduced in the calculation of particle separation in cyclones and a unique analysis for fractional separation efficiency is presented. Calculation of air recirculation is treated in detail with worked examples. Diagrams are elaborated for quick selection of appropriate operational parameters.

5.1 Principles of Dust Separation

The dust transported by the air stream should be separated from the air before it enters the storage bin. The separation can be accomplished by using

- power field (gravitational, centrifugal),
- fabric filtering.

Big and moist sawdust produced in the primary wood processing can well be separated in a gravity chamber, where the air velocity quickly decreases and the particles, due to their gravity, are settling down onto the chamber floor.

Particles of medium sizes, without fine fractions, may be separated using cyclones. Separation of fine fractions requires fabric filters.

Cyclone is an old (around 1880), a simple and reliable separation device that is often used. It consists of an inlet tube, upper cylindrical outlet tube, Fig. 5.1. The air with dust particles enters the upper cylindrical part tangentially producing a rotational motion of air.

The air within the outer wall and central tube moves downward and the velocity field is gradually transforming into a potential flow with a characteristic velocity distribution according to the following equation

E. Magoss et al., *Dust Extraction and Handling in the Wood Industry*,
SpringerBriefs in Applied Sciences and Technology,
https://doi.org/10.1007/978-3-031-08915-2_5

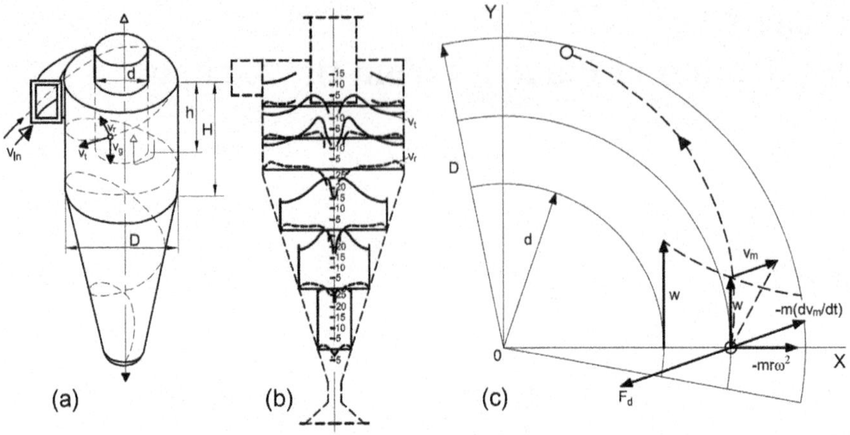

Fig. 5.1 General form of a cyclone and the motion path of particles within the cyclone (**a**), and tangential air velocity distribution in the cyclone (**b**), and velocity and force relations of a moving particle (**c**)

$$r \cdot v_x = \text{constant}$$

which is seen in Fig. 5.1b. After a given downward motion, the air is turning upward and leaves the cyclone through the central tube.

The rotating motion of air causes a centrifugal power field separating the particles from the air. The particles are moving on cycloid path toward the outer wall and slide down into the bottom conical part, Fig. 5.1c.

The motion of air and particles in cyclones is a complicated phenomenon. The air motion in the upper part of the cyclone is more or less similar to a rotating body with constant angular velocity ω. During the downward motion the velocity field will gradually be transformed into a potential field, $r \cdot w_x = \text{constant}$, equalizing the distribution of quantity of motion (momentum). Particles entering the upper part of the cyclone may take arbitrary positions within the free radius and they have different conditions for separation. It is obvious that particles coming near to the outer radii have more chance to arrive at the outer wall.

Another problem is the velocity distribution according to the potential flow. The particle inlet velocity at the inner radii will be less compared to the air velocity but at the outer radii, it will be higher. These problems can only be followed with complicated numerical analysis (Pirumov 1961, 1981). Practical calculations always use different assumptions and simplifications.

The axial velocity in the upper cylindrical part is calculated from the air volume in the unit time and the cross-section as follows

$$v_a = \frac{4 \cdot V}{\pi \cdot \left(D^2 - d^2\right)}$$

The condition for the smallest particle separation is that the particle must arrive the outer wall before the air turns upward into the central tube. We suppose that the forces acting on the particle are the centrifugal force and the drag force, and the equation of motion may be written as (Pirumov 1961, 1981):

$$m \cdot \frac{\mathrm{d}v_r}{\mathrm{d}t} = m \cdot r \cdot \omega^2 - F_d \tag{5.1}$$

where m is the mass of a particle,

F_d is the aerodynamic drag force.

For small particles the radial acceleration is small and, with good approximation, the two forces are in equilibrium

$$m \cdot r \cdot \omega^2 = F_d \tag{5.1a}$$

or

$$m \cdot \frac{v_t^2}{r} = F_d$$

In the literature, by common consent, the Stoke's law valid for laminar flow is used for describing the drag force. It is based on the assumption that the particle Re-number is around 1 or below. This assumption would be correct for particles moving in resting air. But, concerning cyclones with highly turbulent air motion, several contradictions are encountered:

- using Stoke's law, the calculated minimum particle diameters, without any corrections, are in the feasible range,
- according to the measurements, the calculated terminal velocities require, at the same time, fundamental corrections,
- the separation (radial) velocity of particles in a cyclone is directly proportional to the terminal velocity and, therefore, the terminal velocity is fundamental property, its correct use is crucial in calculations.
- using the necessary corrections to obtain real terminal velocities, the calculation with turbulent drag gives also realistic results.

Using Eq. (5.1a), the equilibrium equation reads

$$m \cdot r \cdot \omega^2 = \frac{d^3 \cdot \pi}{6} \cdot \rho_m \cdot r \cdot \omega^2 = F_d$$

and with Stoke's Law

$$v_r = \frac{d^2 \cdot \gamma_m}{18 \cdot \eta} \cdot \frac{r \cdot \omega^2}{g} = v_{cr} \cdot \frac{r \cdot \omega^2}{g}$$

This equation clearly indicates that the particle separation (radial) velocity is directly proportional to the terminal velocity and the ratio of centrifugal and gravitational accelerations. Needless to say that the accuracy of calculations fundamentally depends on the accurate knowledge of terminal velocity.

In the case of constant angular velocity w and taking $v_r = dr/dt$, the above equation is integrated to

$$d_0 = \sqrt{\frac{18 \cdot \eta \cdot v_a}{\rho_m \cdot h \cdot \omega^2} \ln \frac{D}{d}} \tag{5.2}$$

where d_0 is the separated minimum diameter,

v_a is the axial velocity of air in the cyclone,

h is the length of central tube within the cyclone.

The calculation of angular velocity w requires some considerations. The tangential velocity in the cyclone is proportional to the inlet velocity v_{in} and, as a first approximation, we may assume $v_t = v_{\text{in}}$. In this case

$$\omega = \frac{2v_t}{D} = \frac{2v_{\text{in}}}{D}$$

This assumption is, however, the upper bound for the possible value of w and, therefore, we may use a coefficient of safety with a value of 0.85. Accordingly, the effective angular velocity may be calculated as

$$\omega = \frac{1.7 \cdot v_{\text{in}}}{D}$$

Another often-used assumption is the constant tangential velocity distribution. In this case

$$m \cdot r \cdot \omega^2 = m \cdot \frac{v_t^2}{r} = 3\pi \cdot d \cdot \eta \cdot v_r$$

which is integrated to the following final form

$$d_0 = \sqrt{\frac{2.25\eta \cdot v_a \cdot \left(D^2 - d^2\right)}{\rho_m \cdot h \cdot v_t^2}} \tag{5.2a}$$

Due to the tangential inlet, a nearly constant angular velocity evolves in the upper part of the cyclone, and this velocity profile gradually transforms into a potential flow, Fig. 5.2. This transformation is a natural process that equalizes the quantity of motion $(m \cdot v)$ within the space. During this transformation process, the local tangential velocity and herewith the centrifugal acceleration varies which creates different

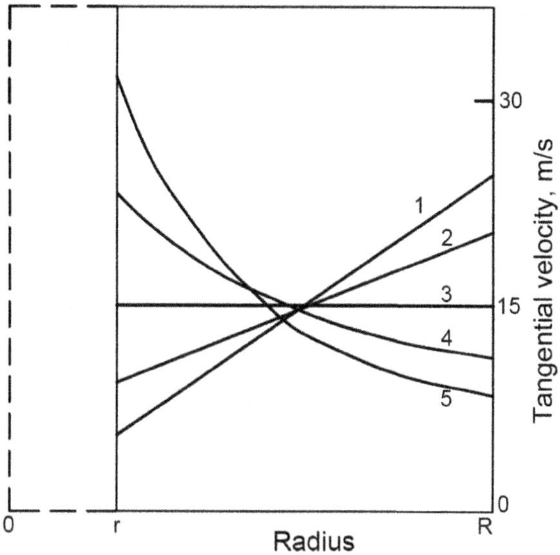

Fig. 5.2 Possible tangential velocity distribution in a cyclone. 1—w = constant, 2—straight, $k = 89$. 3—$v_t =$ constant, 4—$\sqrt{r} \cdot v_t =$ constant, 5—$r \cdot v_t =$ constant. Equal areas under the curves

conditions for particle separation. The effect of different velocity distributions on the separation process has been never examined in detail.

In order to facilitate a general examination, three straight lines with different slopes in Fig. 5.2 can be described with a common equation as a function of slope. Our constraint is, however, that the area under the curves remains constant. The slope of distributions going through a given point is defined as

$$k = \frac{v_t - v_{t0}}{r - r_{av}}$$

where r_{av} is the average radius belonging to the chosen average tangential velocity v_{t0} in the intersection point.

From this equation we get

$$v_t = v_{t0} + k(r - r_{av})$$

In our example, $v_{t0} = 15$ m/s and $r_{av} = 0.09375$ m. Note that in this case the angular velocity w is not constant but increases inwards from 133 (outer wall) to 266 (curve No. 2). At the same time, the tangential velocity varies from 20 to 10 m/s and the corresponding slope is $k = 88.88$. In our case, if $k = 160$, the distribution is equal to curve No. 1 with constant angular velocity of 160.

The equilibrium equation supplies the following equality

$$\frac{d^2 \cdot \rho_m \cdot h}{18 \cdot \eta \cdot v_a} = \int_r^R \frac{r \cdot \mathrm{d}r}{[v_{t0} + k(r - r_{\mathrm{av}})]^2} = I(k)$$

and

$$d_0 = \sqrt{\frac{18 \cdot \eta \cdot v_a \cdot I(k)}{\rho_m \cdot h}} \tag{5.2b}$$

For the case of constant angular velocity, $k = k_{\max}$, both Eqs. (5.2) and (5.2b) give exactly the same result. In Eq. (5.2b) the integral is numerically calculated.

Similarly, the curves No. 3, No. 4, and No. 5 can be described by a common equation

$$r^n \cdot v_t = \text{constant}$$

where the exponent n varies between 0 and 1. The constant is given by the intersection point in Fig. 5.2 as

$$\text{constant} = r_{\mathrm{av}}^n \cdot v_{t0}$$

where r_{av} is the average radius belonging to the chosen average tangential velocity v_{t0}. In this case, the general equation has the following form:

$$d_0 = \sqrt{\frac{18 \cdot \eta \cdot v_a \cdot \left(R^{2n+2} - r^{2n+2}\right)}{\rho_m \cdot h \cdot \left(r_{\mathrm{av}}^n \cdot v_{t0}\right)^2 \cdot (2n+2)}} \tag{5.2c}$$

This last equation is more convenient to perform detailed analysis as a function of exponent n.

Example. We consider a cyclone with the following input data: $D = 0.3$ m, $d = 0.075$ m, $v_{\mathrm{in}} = 20$ m/s (inlet cross-section 6×5.53 cm), $h = 0.18$ m, the air volume rate $V = 239$ m^3/h $= 0.0664$ m^3/s, the entering quantity of motion is 1.62 Ns/s, $v_a = 1$ m/s, and $r_{\mathrm{m}} = 400$ kg/m^3. Using the derived equations, for the typical distributions we obtain the following minimum particle diameter given below

Curve No.	d_0, μm	Type of flow
1	15.6	w = constant
2	14.2	
3	14.52	v_t = constant
4	15.7	

(continued)

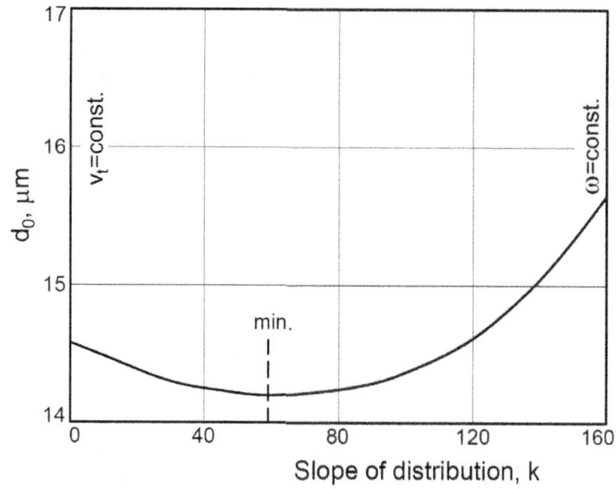

Fig. 5.3 Influence of the slope of the tangential velocity distribution on the separated minimum particle diameter

(continued)

Curve No.	d_0, μm	Type of flow
5	17.64	$r \cdot v = \text{constant}$

The minimum was obtained at a slope of $k = 60$ (Fig. 5.3) with a value of 14.2 μm (a similar case to curve No. 2), but almost the same obtained with constant tangential velocity, 14.5 μm. The constant angular velocity (curve No. 1) and the potential distributions gave somewhat worse results. It can however be stated that the examined distributions, especially the first four, do not differ very much from each other. The fully developed potential flow is not very common in practice.

The obtained results are in the feasible range without correction. Two fundamental errors are made in these calculations but, fortunately, they are on the opposite sides: the use of Stoke's law causes an over-estimated drag force which is counterbalanced by an over-estimated mass force.

In the following, we use the turbulent drag force with proper correction based on our measurement results given in Sect. 1.6.

We consider now a wood particle of 20 μm diameter. Its terminal velocity is accurately not known but an extrapolation of our measurements makes values between 0.015 and 0.04 m/s probable. Using the higher value, the necessary correction of the theoretical value can be calculated. Due to the small size, the irregular shape and hairy surface, the mass-related diameter (d_m) will be smaller, and the blown face-related diameter (d_a) even may be higher compared to the screen size d_0. Therefore we take

$$d_m = k_1 \cdot d_0 \quad \text{and} \quad d_A = k_2 \cdot d_0$$

Using Eq. (1.9), the terminal velocity equation, compared to Eq. (1.10), modifies to

$$v_{cr}^{+} = \sqrt{\frac{k_1^3}{k_2^2}} \cdot v_{cr}$$

If we take $\gamma_m = 4000$ N/m^3 and $c_e = 1.0$, then the theoretical terminal velocity is 0.297 m/s which, compared to the real value of 0.04 m/s, requires a correction of $\sqrt{k_1^3/k_2^2} = 0.1346$. A combination of $k_1 = 0.33$ and $k_2 = 1.3$ gives the required correction.

Now the equilibrium equation for turbulent flow reads

$$m \cdot r \cdot \omega^2 = m \cdot \frac{v_t^2}{r} = c_e \cdot A \cdot \frac{\rho_0}{2} \cdot v_r^2 \quad \text{with} \quad v_t = \text{constant}$$

from which the radial velocity of motion has the following expression

$$v_r = \sqrt{k_1^3/k_2^2} \cdot v_{cr} \cdot \frac{v_t}{\sqrt{r \cdot g}}$$

Again the radial (separation) velocity depends on the terminal velocity and the square root taken from the ratio of field accelerations.

Taking $v_r = \mathrm{d}r/\mathrm{d}t$ and $v_t = \text{constant}$, after integration we get

$$d_0 = \frac{k_2^2 \cdot v_a^2 \cdot c_e \cdot \rho_0}{k_1^3 \cdot h^2 \cdot v_t^2 \cdot 3\rho_m} \cdot \left[R^{1.5} - r^{1.5}\right]^2 \tag{2d}$$

Using the data of our earlier example ($v_t = 15$ m/s), the above equation gives $d_0 = 16.8$ μm for the separated particle entering at the inner radius. It can be seen that the turbulent drag supplies full realistic value if the proper correction is taken into account.

Now we follow the formerly experienced way to get general equations for the typical velocity distributions. Concerning first the three linear distributions, the equilibrium equation reads

$$m \cdot \frac{v_t^2}{r} = \frac{(k_1 \cdot d_0)^3 \cdot \pi \cdot \rho_m}{6} \cdot \frac{(v_{t0} + k(r - r_{av}))^2}{r}$$
$$= c_e \cdot \frac{(k_2 \cdot d_0)^2 \cdot \pi \cdot \rho_0}{4.2} \cdot v_r^2$$

The time available to separation is $t = h/v_a$ and the above equation is brought to the form:

$$\sqrt{\frac{4 \cdot k_1^3 \cdot \rho_m \cdot d_0}{3 \cdot k_2^2 \cdot c_e \cdot \rho_0}} \cdot \frac{h}{v_a} = \int_r^R \frac{\sqrt{r} \cdot \mathrm{d}r}{v_{t0} + k(r - r_{\mathrm{av}})} = I(k)$$

and finally

$$d_0 = \frac{3 \cdot k_2^2 \cdot v_a^2 \cdot c_e \cdot \rho_0}{4 \cdot k_1^3 \cdot h^2 \cdot \rho_m} \cdot I(k)^2 \tag{2e}$$

where the integral is numerically calculated. If $k = 0$, then v_t = constant and Eqs. (5.2d) and (5.2e) supply the same result.

The curves No. 3, 4, and 5 can also be treated with a common equation

$$r^n \cdot v_t = \text{constant} = r_{\text{av}}^n \cdot v_{t0}$$

The equilibrium equation has now the following form:

$$m \cdot \frac{v_t^2}{r} = \frac{(k_1 \cdot d_0)^3 \cdot \pi \cdot \rho_m}{6} \cdot \frac{r_{\text{av}}^{2n} \cdot v_{t0}^2}{r^{2n+1}}$$
$$= c_e \cdot \frac{(k_2 \cdot d_0)^2 \cdot \pi \cdot \rho_0}{4.2} \cdot v_r^2$$

which is integrated to the following final form:

$$d_0 = \frac{3 \cdot k_2^2 \cdot v_a^2 \cdot c_e \cdot \rho_0}{k_1^3 \cdot h^2 \cdot v_{t0}^2 \cdot r_{\text{av}}^{2n} \cdot \rho_m} \cdot \left(\frac{R^{\frac{2n+3}{2}} - r^{\frac{2n+3}{2}}}{2n+3} \right)^2 \tag{2f}$$

Using the data given in our earlier example, for turbulent drag the following values are obtained.

Curve No.	d_0, μm	Type of flow
1	19.5	w = constant
2	16.9	
3	16.3	v_t = constant
4	17.5	
5	19.1	$r \cdot v$ = constant

At the first glance, the picture is changed not too much. The influence of tangential velocity distribution on the separable particle diameter is also moderate. At the same time, this calculation method is physically well established and the correction can always be chosen to the given conditions.

It should be stressed that a proper correction can only be calculated in the knowledge of experimentally determined terminal velocities. Our knowledge is, however, very incomplete and efforts should be made for filling this gap.

The derived equations for the separation (radial) velocity have clearly indicated the influence of acceleration field on the efficiency of cyclones. The power field of cyclones can be characterized by the ratio of centrifugal and gravitational accelerations

$$\mathrm{PF} = \frac{m \cdot v_t^2 / r}{m \cdot g} = \frac{v_t^2}{g \cdot r} \tag{5.3}$$

where r is the radius of the cyclone. Seemingly, the increase of tangential velocity is a good choice, but it is associated with higher pressure drop which, apart from higher energy consumption, may be a limiting factor concerning system stability for a given ventilator. If high separation efficiency is required, the practice selects smaller radius and more units (multi-cyclones).

The pressure loss in cyclones is determined similarly to suction hoods and bent tubes (see in Sect. 4.3):

$$\Delta p_c = \xi_c \cdot \frac{\rho}{2} \cdot v_{\mathrm{in}}^2$$

and the loss factor ξ_c generally will be measured and given by the manufacturer for air (idling). It is interesting that the pressure loss is smaller under material load. The reducing effect of dust particles on the pressure drop may be explained by the turbulizing effect of moving particles in the laminar boundary layer in the close vicinity of the wall. This turbulizing effect depends, however, on particle size and, therefore, the higher portion of fine particles decreases the pressure drop reduction.

As we have seen above, cyclones can not separate the fein fraction under 10–15 μm. In order to separate also the fein fraction, *fabric filters* are used. They may be used as primary or secondary separation devices. In the latter case, the fabric filter follows a cyclone as a pre-separator.

The general layout of a fabric filter using mechanical shaking mechanism for cleaning is shown in Fig. 5.4. The long cylindrical bags ensure a high surface per volume ratio and hereby a low flow velocity through the fabric. The dust-laden air enters the filter from beneath and so the larger particles may be separated by gravity falling direct into the hopper. Fine particles are moving between the bags toward the fabric surface where they are separated.

The air to fabric surface ratio is defined as the ratio of air volume rate divided by the area of filter fabric and it gives the flow velocity of air through the filter fabric related to the surface area of the filter, For fein particles, its typical value is around 1.5 m/min and for coarser materials, it may be doubled. Its optimum value depends, however, on the dust concentration and the average particle size. Increasing concentration and smaller particle size require smaller face velocity. Using dilute phase transport in the wood industry, an approximate face velocity as a function of median size of particles is given by the following empirical equation:

$$v_f = 0.3 \cdot \ln(10 \cdot d_m), \quad \mathrm{m/min}$$

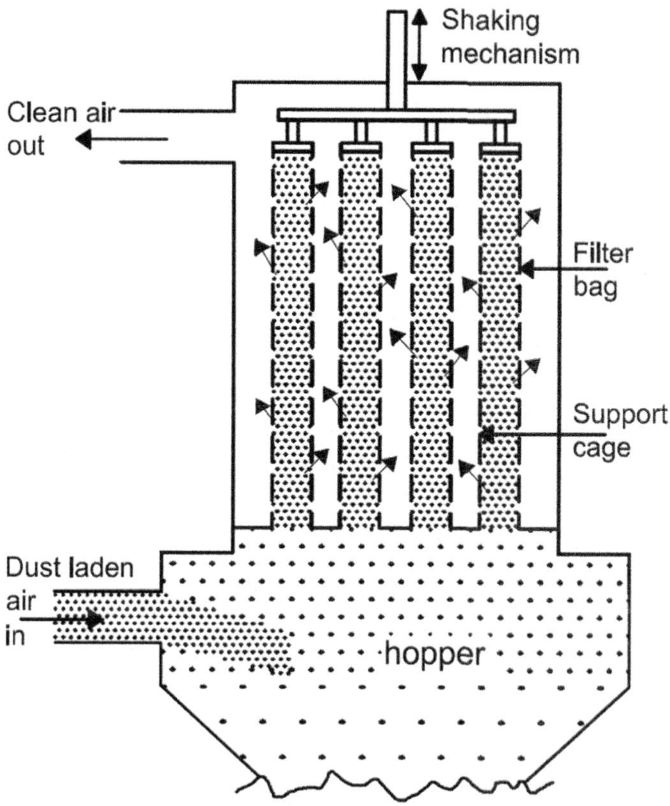

Fig. 5.4 General layout of a fabric filter using mechanical shaking mechanism for cleaning

where the median diameter must be substituted in Micron.

The process of filtration on a fabric filter is a complex one. In the first time, a sieving occurs by the interstices of the filter fabrics. Very fine particles are also caught by impingement on the fibers within the filter fabric. Particles too large to press through the mesh are retained on the surface of the filter and they build-up and form a caked layer on the surface with a given porosity. The airflow path through the cake has a labirinth nature with increasing pressure drop while the effective mesh size decreases, the filtering action increases. Figure 5.5 demonstrates experimental results showing the porosity change in the caked layer for different pressure drops (Schmidt 1993). As the layer thickness increases, the porosity of the cake toward the fabric decreases. A greater pressure presses the fine particles deeper into the cake and further decreases the porosity. The increasing thickness and decreasing porosity always mean an increasing pressure drop through the cake. As a consequence, the pneumatic conveying system will be influenced by varying pressure drops of the fabric filter.

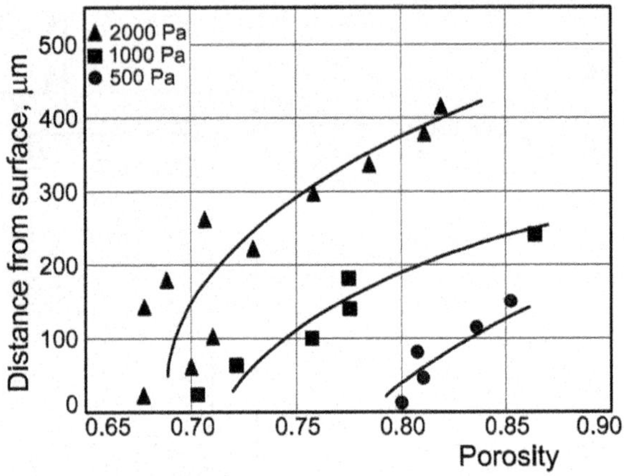

Fig. 5.5 Porosity of caked layer on the surface of the filter for different pressure losses as a function of distance from the fabric surface (Schmidt 1993)

In order to avoid the blockage of the system, the fabric filter must be cleaned either periodically or continuously. The filter unit has several compartments and one compartment can be shut off for cleaning while the others have remained in operation. It is an important aspect that the proportion of fabric area out of action would be minimized which ensures a lower pressure drop variation in the entire system.

The cleaning action can be accomplished using mechanical shaking or air pulsing. Shaking the framework induces inertia forces resulting in the separation of caked layer from the fabric surface and the dislodged material is falling into the bottom of collecting hopper (see Fig. 5.4).

The air resistance of fabric filters for clean air is generally low and amounts to less than 10 N/m^2. Due to the low flow velocity through the fabric, the air resistance is proportional with the velocity and it can be calculated with the simple equation as

$$\Delta p_{f0} = K_f \cdot \eta \cdot v_f \quad \text{N/m}^2 \tag{5.4}$$

where η is the air viscosity (18.3×10^{-6} Ns/m^2 at 20 °C),

K_f is a constant determined experimentally.

Its value for cotton, wool, orlon, and teflon fabrics varies between 3×10^6 and 7×10^6 1/m. The face velocity must be substituted in m/s.

The calculation of resistance to dust-laden air as a function of working time is much more uncertain. The bigger particles losing their velocity are separated due to gravity and falling down into the hopper (see Fig. 5.4).

The finer particles reach the fabric surface or its vicinity and they either build-up a cake on the fabric surface or fall down. This process is highly dependent on the surface property of particles and their size distribution.

Concerning wood particles, it is likely that the quite fine particles have good adhering properties promoting the build-up of a cake. Sadly, there is no experimented material about this question that would clear this problem in detail. Some considerations may, however, be done.

It may be assumed that the air resistance is proportional to the cake thickness which is time-dependent. If we assume that all dust will be separated by the fabric and remained on the fabric surface, the cake thickness is calculated as

$$h(t) = \frac{\mu_t \cdot \rho_0 \cdot \varepsilon \cdot v_f \cdot t}{\rho_m}, \quad \text{m} \tag{5.5}$$

where μ_t is the dust transport concentration,
ρ_0, ρ_m are the air and material density, respectively,
ε is the dust and solid wood ratio (2.5–3.0),
t is the elapsed time.

Evaluation of Eq. (5.5) shows that, with this assumption, the fabric surface would very quickly be laden with a thick dust layer. For example, with $\mu_t = 0.5$, $\varepsilon = 2.5$, $\rho_m = 500$ kg/m^3, $v_f = 3$ m/min and $t = 1$ min the dust thickness would be $h = 9.1$ mm. In practice, it is far not the case. The reason is, as above explained, that only the fine fraction will build-up the cake on the fabric surface.

The pressure drop on the dust-laden fabric can be calculated with an equation similar to Eq. (5.4) in the following manner:

$$\Delta p_f = K_d \cdot \eta \cdot h(t) \cdot v_f, \quad \text{N/m}^2 \tag{5.6}$$

where K_d is an experimentally determined constant which takes into account the relative proportion of fine fraction in the dust and its adhering properties. Producing air-dry dust and depending on the particle size distribution, the following advice may be done: if the fine fraction under 15 μm is less than 1% then $K_d = 1.5 \times 10^7$ 1/m^2. Separating pure sanding dust, the fine fraction may be 3–4%, and in this case, K_d will be multiplied by the percent amount of fine fraction. Using the data of above example with $t = 30$ min, $\Delta p_f = 184$ N/m^2. Using sanding dust for longer time, the pressure drop may increase considerably. Equations (5.5) and (5.6) indicate that the pressure drop more or less linearly increases as a function of working time.

After cleaning the fabric filter, the pressure drop decreases to a value somewhat higher than that of the new clean fabric. Some fein particles adhered to the fibers within the fabric will not be removed by the cleaning process. Therefore, the pressure drop pattern as a function of working time will be such depicted in Fig. 5.6.

5.2 Separation Efficiency

One of the most important characteristics of separation devices is their separation efficiency. The initial dust has a given particle size distribution (see in Sect. 1.2) and, after separation, the passed and retained parts have also some given distributions. That means that the separation device cannot make a sharp size separation.

Two different separation efficiencies may be defined:

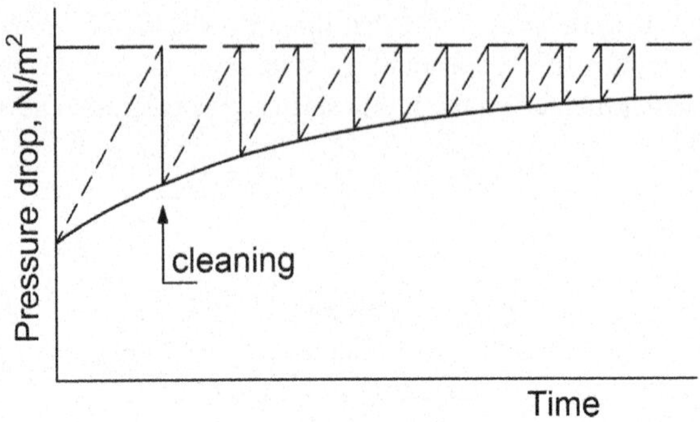

Fig. 5.6 Cyclic variation of pressure drop of a fabric filter due to periodic cleaning

- overall efficiency which is the ratio of retained mass and the total (inlet) mass,
- fractional efficiency related to a narrow fraction and it varies as a function of particle size.

The overall separation efficiency is determined by the following relation

$$\eta = \frac{G_{\text{in}} - G_{\text{out}}}{G_{\text{in}}} \tag{5.7}$$

where G_{in} is the inlet mass,

G_{out} is the passed (outlet) mass.

It is obvious that the overall efficiency is characterized by a number. The fractional efficiency is calculated in the same way as before, but for a narrow size fraction

$$\eta_{fr} = \frac{\Delta G_{\text{in}} - \Delta G_{\text{out}}}{\Delta G_{\text{in}}} \tag{5.8}$$

If the size fractions are sufficiently narrow then the sequence of single values describes a curve valid for the whole size distribution curves as shown in Fig. 5.7 (Sitkei 1994). The curves are asymptotic and can be described with exponential function in the following form:

$$\eta_{fr} = 100 \cdot \left[1 - e^{-\left(\frac{x}{X_k}\right)^n}\right] \quad \% \tag{5.9}$$

where X_k is the characteristic particle size at which the separation efficiency is 63.21%,

n is an exponent adjusting the curve to the experimental results.

Modifying the above two parameters, Eq. (5.9) can well fit the experimentally determined curve.

It is important that, concerning a given filter device, the fractional separation efficiency is an independent characteristic while the overall separation efficiency

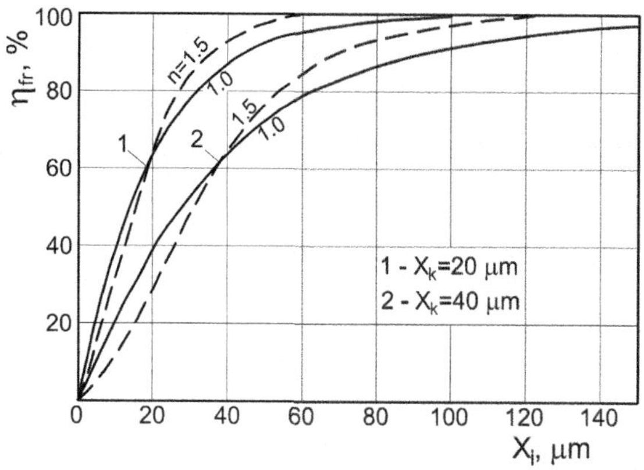

Fig. 5.7 Typical fractional separation curves

depends also on the particle size distribution of dust:

$$\Delta Q = f\left(\eta_{fr}, \overline{X}, \sigma\right)$$

5.3 An Analysis of Fractional Separation Efficiency

Analytical description of particle size distribution and fractional separation efficiency allows to perform detailed analysis highly facilitating the quick selection of separating devices for a prescribed separation efficiency. On the other hand, the performance of existing system can reliably be judged using this evaluation method.

The dust quantity passed the filter can be given as

$$\Delta Q = (1 - \eta) \quad \text{decimal}$$

or

$$\Delta Q = \sum_{x=0}^{x_{\max}} \left(1 - \eta_{fr}\right) \cdot y.$$

Using Eqs. (1.2) and (5.9), the above equation takes the following more detailed form (Sitkei 1994)

$$\Delta Q = \frac{100}{\sqrt{2\pi} \cdot \overline{X} \cdot \sigma \cdot e^{\sigma^2/2}} \int_0^{x_{\max}} \exp\left\{-\left[\left(\ln x/\overline{X}\right)^2/2\sigma^2 + (x/X_k)^n\right]\right\} \mathrm{d}x, \quad \% \tag{5.10}$$

Systematic numerical evaluation of Eq. (5.10) allows to construct diagrams that serve for quick estimation. It was recognized that the $X_k/\overline{X}$ ratio uniquely determines the dust quantity ΔQ passed through the filter and pollutes the environment. Further variables are the standard deviation σ and the exponent n.

The evaluation of Eq. (5.10) is demonstrated in Fig. 5.8 (Sitkei 1994). An effective separation is possible for $X_k/\overline{X}$ ratios less than 0.1. For higher requirements ($\eta \geq 99\%$) the required ratio is around 0.05 and the standard deviation of size distribution has considerable influence on the achievable efficiency.

The exponent n in Eq. (5.9) has also considerable effect on the dust quantity ΔQ, especially at high efficiencies, Fig. 5.9. For the $X_k/\overline{X} = 0.05$ the variation may be threefold or even more.

The size distribution of dust passed the filter may be interesting concerning health risks. This calculation method is suitable to follow this problem. The results of such calculations may be summarized as follow

- the size distribution of the ΔQ quantity is not accurately logarithmic normal but similar to that,
- the new modus $\overline{X}'$ is much less compared to the initial value $\overline{X}$ and its value is uniquely determined by the $X_k/\overline{X}$ ratio, Fig. 5.10.

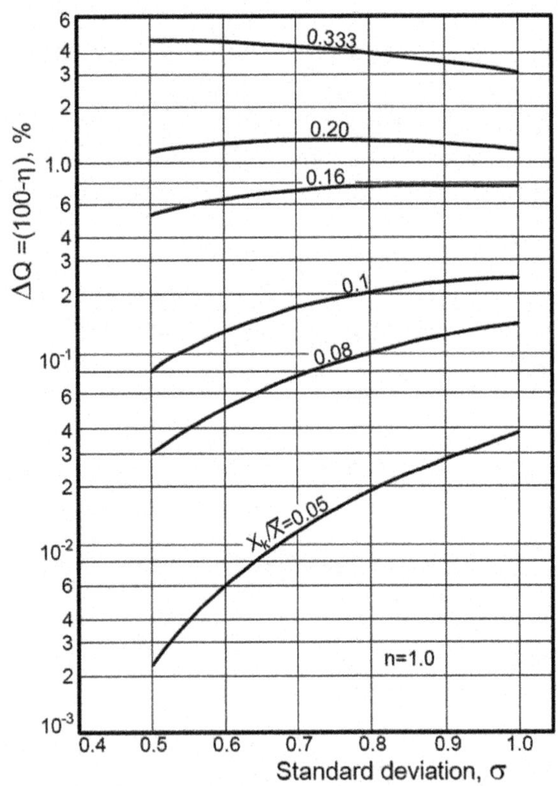

Fig. 5.8 Dust quantity passed the filter as a function of standard deviation for different $X_k/\overline{X}$ ratios

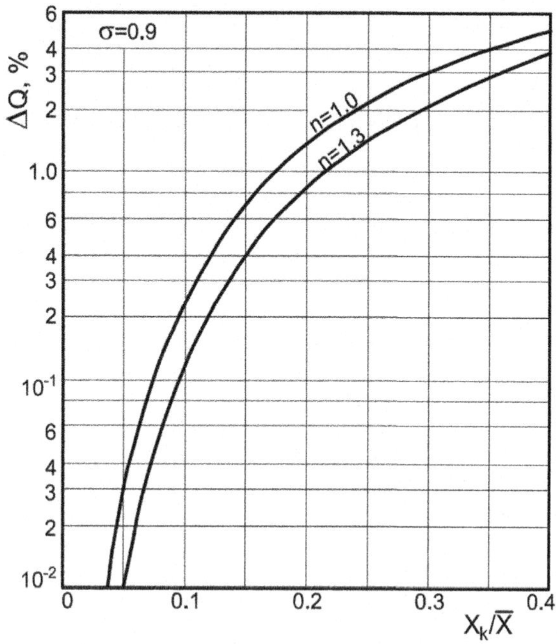

Fig. 5.9 Effect of exponent n on the dust quantity passed the filter

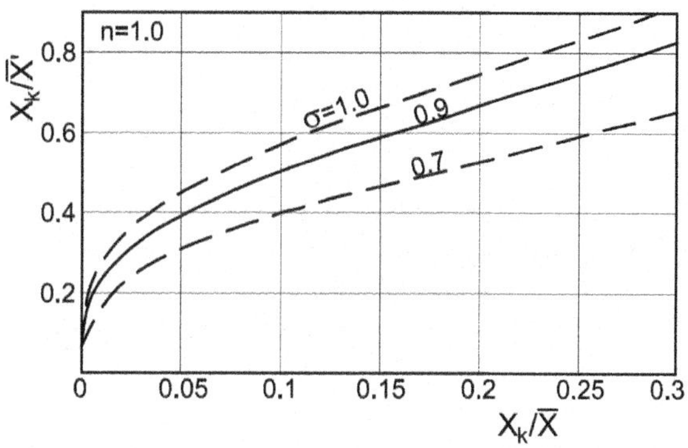

Fig. 5.10 Auxiliary diagram for determining the modus of dust distribution passed the filter with known X_k values

- the standard deviation of new distribution is also decreased compared to the initial value. In these calculations standard deviations of 0.55–0.65 are obtained.

The curves given in Fig. 5.10 can empirically be described by the following equation

$$\frac{X_k}{\overline{X}'} = 1.44 \cdot \left(X_k/\overline{X}\right)^{0.4} \cdot \sigma \tag{5.11}$$

Example. A circular saw produced the following dust characteristics: $\overline{X} = 0.09$ mm, $Mi = 0.23$ mm, $X_{av} = 0.36$ mm, $\sigma = 0.97$ and $\eta = 99\%$ is required. From Fig. 5.6 $\frac{X_k}{\overline{X}} = 0.18$, $X_k = 16.2$ μm ($n = 1.0$). The new modus is $\overline{X}' = 22.8$ μm compared to the initial value of 90 μm. For this purpose a multi-cyclone is appropriate.

For higher requirements with $\eta = 99.9\%$, the $\frac{X_k}{\overline{X}} = 0.07$ value can be read from Fig. 5.8 and $X_k = 6.3$ μm. The modus $\overline{X}'$ of the dust passed the filter is 12.8 μm, using Fig. 5.10. A fabric filter should be selected.

5.4 Air Recirculation

Wood processing operations almost always pollute the environment in a certain extent. The mechanical woodworking is associated with dust and chip formation and the fein particles may enter the human respiratory organs causing serious health risks. Especially particles around 5 μm and below can float long time in the air and will be breathed by the workers. Depending on the particle size, they get through the nose and the throat into the lung. Furthermore, workers need, even without artificial pollution, fresh air to maintain health environment. In order to ensure clean and fresh air in the workroom, continuous ventilation of the workroom is needed.

The dust and chips produced by woodworking machines will be captured and removed by air stream and this dust-laden air leaves the workroom which should be replaced with clean air. Further problems may arise in hot or cool weather when the external air must be heated or cooled before entering the workroom, which requires additional energy. In this case, the recirculation of filtered air may lower the energy costs.

The schematic layout of filtered air recirculation is shown in Fig. 5.11. The dust-laden air sucked from the woodworking machines is transported into the filter. The filtered air is recirculated, fully or partly, into the workroom.

The dust load at the working place is limited by national and international standards, for example, 3 mg/m^3 in Hungary.

This is a strong requirement fulfillment of which is possible only by careful design and selection of filter equipment. The calculation of expected dust concentration in the workroom is not free from uncertainties, especially concerning the quantity of dust escaping from the woodworking machines direct into the workroom.

The layout of a ventilation system with air recirculation is depicted in Fig. 5.12. Pollution of air in the workroom is originated from dust escaped from the woodworking machines and dust contained in the recirculating air.

The air entering the room is equal the air leaving the room

$$m_v + m_r = m_0 + m_{\text{out}}$$

where m_v is the ventilating air, kg/s.

m_r is the recirculating air ($m_r = R \cdot m_0$), kg/s.

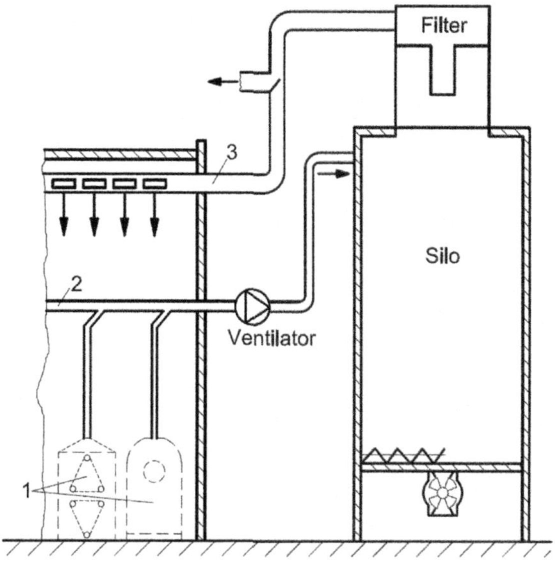

Fig. 5.11 Recirculation of filtered air into the workroom. 1—Woodworking machines, 2—main duct, 3—recirculated air

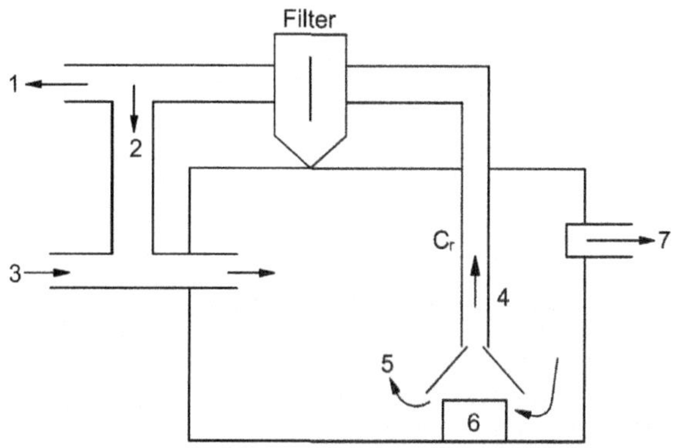

Fig. 5.12 Ventilation and filtered air recirculation in a workroom with dust source and dust suction. 1—filtered air outlet, 2—recirculated air, 3—clean air inlet, 4—sucked air from the dust source, 5—dust escaping into the workroom, 6—dust source, 7—outlet of ventilating air

m_0 is the air sucked from the machines, kg/s.
m_{out} is the outlet air, kg/s.
From the produced dust m_a a small portion pollutes the workroom

$$m_{a1} = (1 - \eta_s) \cdot m_a, \quad \text{kg/s}$$

where η_s is the efficiency of dust removal by the suction air. The dust-laden air with $\eta_s \cdot m_a$ dust content enters the filter and after filtering, it will be reduced to a value of

$$(1 - \eta) \cdot \eta_s \cdot m_a, \quad \text{kg/s}$$

which is partly getting into the external environment and partly recirculated into the workroom.

The sum of dust leaving the workroom is

$$(m_0 + m_{\text{out}}) \cdot C_r$$

where C_r is the average dust concentration in the workroom, kg dust/kg air.

The equality of introduced and removed dust yields

$$C_r = \frac{m_a \cdot [(1-\eta_s) + R \cdot \eta_s \cdot (1-\eta)]}{m_0 + m_{\text{out}}} \tag{5.12}$$

where R is the portion of recirculated air.

From Eq. (5.12) it is obvious that, in the interest of low dust concentration in the workroom, high-efficiency suction hoods and dust filters, and abundant ventilating air are required.

The appropriate selection of the efficiency η_s requires some considerations. Bigger particles not removed by the suction air are settled by gravity on the near surface and do not pollute the workroom. Only small particles able to float longer time in the air will be spread in the workroom. Particle sizes of 5 μm and below are floating freely in stationary air and, therefore, their portion in the dust is important information. Figure 5.13 shows measurement results for dust produced by sanding and bandsaw. Taking particle sizes 5 μm and below, the portion is 3×10^{-4} and 1.8×10^{-3}, respectively. That is the total portion much of which will be removed by the suction air. Some ten percent of this fine fraction may get into the workroom but, sadly, no reliable data are available.

Example. Ten woodworking machines are operating in workroom, the total dust production is 0.5 kg/s. The total suction air is 10,000 m^3/h = 2.778 m^3/s = 2.27 kg/s. Fein fraction getting into the air is $0.1 \times 3 \times 10^{-4} \times 0.5 = 1.5 \times 10^{-5}$ kg/s, filter efficiency is $\eta = 99.99\%$, ventilating air is 40,000 m^3/h = 11.11 m^3/s = 9.1 kg/s. Portion of recirculation is $R = 0.5$. The average dust concentration, using Eq. (5.12), is 5.2 mg/m^3. Without recirculation the dust concentration is 1.2 mg/m^3.

It may be concluded that the use of recirculation requires a filter with high separation efficiency. Using a pre-separation, the dust load on the fine filter can be considerably reduced which may lower the dust passing the filter in a certain extent.

In order to achieve the filter efficiency of 99.99% (0.01% dust passing the filter), the characteristic number $X_k/\overline{X}$ should be around 0.04 (see in Sect. 5.3). For knife machining the modus $\overline{X}$ is 100 μm or higher and the required filter characteristics X_k should be around 4 μm. Using sanding machines, $\overline{X} \cong 30$ μm and $X_k = 1.2$ μm.

Workrooms in the wood industry are generally polluted in various extent and they should properly be ventilated with clean air. Under cool or hot climate the external air should be heated or cooled which requires equipments and considerable energy. The capacity of equipments and the consumed energy can be decreased by using air

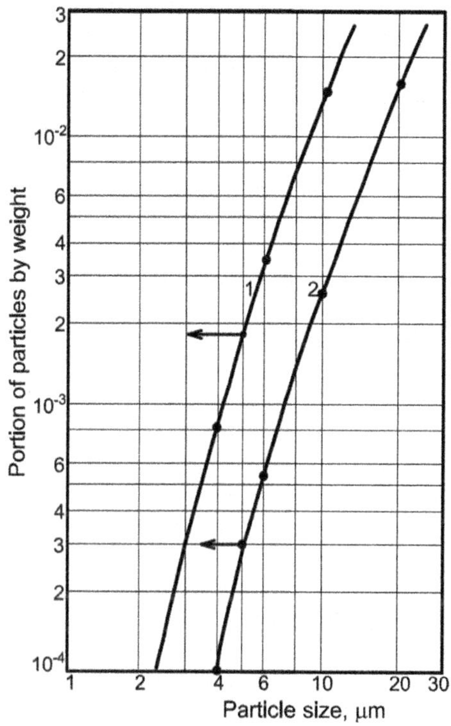

Fig. 5.13 Portion of fine particles by weight as a function of particle size. 1—Dust made by sanding with P-120, 2—dust made by bandsaw (Beechwood)

recirculation. In the latter case the outlet air of the filter unit, in full or partial extent, will be brought back to the workroom as a part of ventilation.

It is clear that the whole amount of ventilating air can not be the recirculating air. Recirculating air contains fein dust particles which in the next filtering process almost will not be separated, continuously increasing the dust concentration in the ventilating air. Therefore, the portion of recirculating air highly depends on the efficiency of separation devices and the transport concentration of dust.

The strong dependence of dust concentration as a function of portion of recirculation and filter efficiency is depicted in Fig. 5.14. The best and cheapest method is the accurate collecting and removing of dust from the place of origin.

The dust concentration of air passed the filter is calculated as

$$c_d = 10^6 \cdot \mu_t \cdot \rho_0 \cdot (1 - \eta), \quad \text{mg/m}^3 \tag{5.13}$$

where the transport concentration μ_t and the separation efficiency η must be substituted in decimal and the air density ρ_0 in kg/m^3. In order to use Eq. (5.13), the overall separation efficiency η can be estimated from Fig. 5.6. For higher transport concentration a high degree of separation is required.

The dust load at the working place is limited by national and international standards, for example, 3 mg/m^3. Generally, this value can be ensured only by dilution

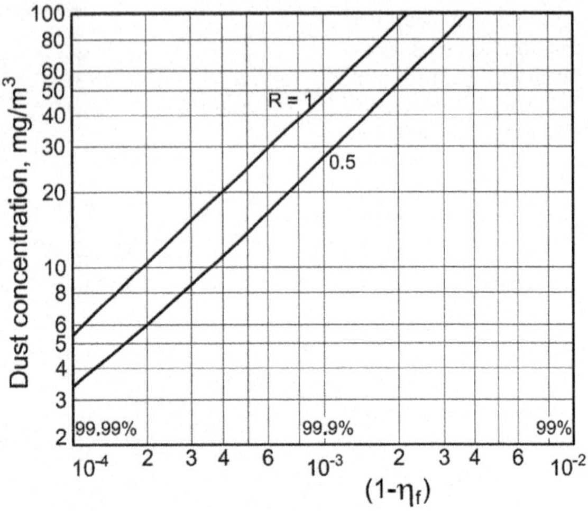

Fig. 5.14 Dust concentration in a workroom using recirculation as a function of filter efficiency

of recirculated air with external clean air. If we assume that a perfect mixing of the recirculated and clean air occurs then the required mixing ratio is simply calculated as

$$\frac{V_r}{V_{ex}} = \frac{c_a}{c_d - c_a} \quad \text{or} \quad V_{\text{ex}} = \frac{c_d - c_a}{c_a} \cdot V_r \tag{5.14}$$

where V_r, V_{ex} are the recirculated and external air, respectively in m^3/s,
c_a is the allowable dust concentration in mg/m^3.

Example. The following input data are used: the overall separation efficiency is 99.99%, the dust concentration in extraction tubes is 0.1 and 0.3, the portion of fein fraction is 10^{-3} (Fig. 5.11), 10% of this fein fraction is getting out to workroom. The calculation results are summarized in Fig. 5.15. The most important influencing factors are the dust concentration μ_{t} and ventilating air to suction air ratio m_{v}/m_0. The portion of recirculated air R has also definite effect on the achievable dust concentration in workroom. In order to reduce the input dust concentration of fein filter, the use of a pre-separation (cyclone) may be the solution.

The curves given in Fig. 5.15 can be described by the following empirical equation

$$C_r = \frac{\mu_t(67 + 31.5R)}{(m_v/m_0)^{1-0.33R}}$$

Note that the numerical values in this equation are only valid for input data given above.

A perfect mixing of fresh and recirculated air can, however, rarely be achieved, but it is not always required. An appropriate introduction of recirculated and clean air into the workroom may solve the problem- The close vicinity of working places should be ventilated by clean air and the recirculated air is introduced far-off the

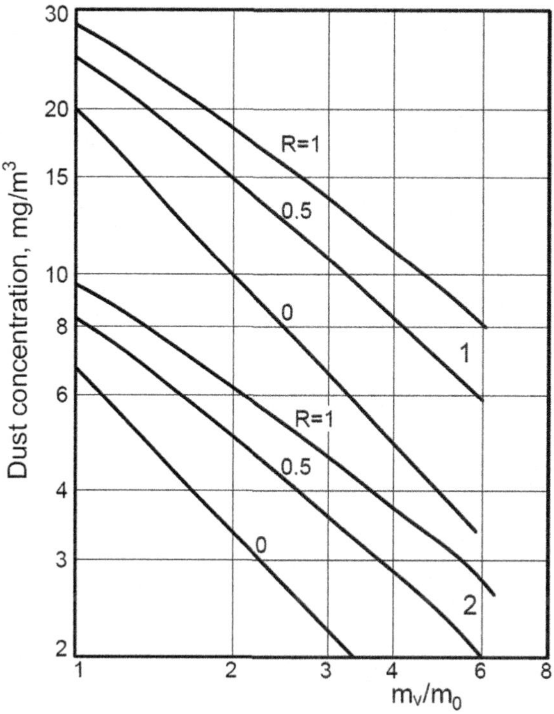

Fig. 5.15 Dust concentration in workroom with ventilation and air recirculation. 1—$\mu_t = 0.3$, 2—$\mu_t = 0.1$

workplaces. If a good mixing is required, the use of a common mixing tube may be the solution. For the sake of safety, the volume of clean air may be increased with some 20% compared to Eq. (5.14).

The wood moisture content may have an interesting influence on the workplace pollution. At higher moisture contents, the fein wood particles are inclined to coagulation, i.e., the small particles are adhered to the bigger ones. Measurements have shown that at a CNC workplace the respirable dust concentration is considerably decreased for 18% moisture content, Fig. 5.16. The total dust concentration, however, is practically remained similar to that of lower moisture content.

As it is shown in the previous section, the size distribution of dust in the recirculating air can also be estimated.

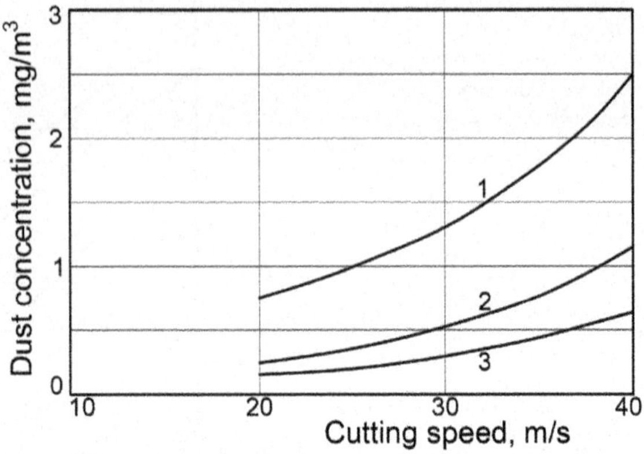

Fig. 5.16 Total and respirable dust concentration at a CNC workplace. Oakwood, milling tool of 60 mm dia. 1—Total, moisture content is 10–18%, 2—respirable, 10% m.c. 3—respirable, 18% m.c.

References

E. Schmidt, Zur Kompression von auf Filtermedien abgeschiedenen Staubschichten (Staub, 1993), pp. 369–376

G. Sitkei, Faipari műveletek elmélete (Theory of Wood Processing) (Budapest, 1994), pp. 485–507

A. Пирумов, Аэродинамические основы инерционной сепарации (Aerodynamics of Inertial Separation) (Госстройиздат, 1961)

A. Пирумов, Обеспыливание воздуха (Dust Reduction of Air) (Машгиз, 1981)

Chapter 6
Auxiliary Equipments

In order to operate an extraction and conveying system, some important auxiliary equipment's are needed. One of the most important such an equipment is the ventilator (prime mover) which provides the necessary energy to the conveying gas. In the design of an extraction and conveying system, it is a need to identify both the gas flow rate and the pressure drop to ensure reliable transportation.

To introduce solid particles into the flowing air stream, feeding devices are needed. In dust extraction systems a very special feeding device is the machine hood in which the tool throwing effect and the developing air velocity field together feed the particles into the pipeline. In other cases, rotary valves and screw feeders are used. In the latter case, the particles are almost at rest and they need to be accelerated to some steady conveying velocity. This means a change in momentum requiring some additional energy.

The arrangement of pipelines is highly determined by the number and location of woodworking machines within the workroom and, therefore, quite different pathways are possible. An optimum design, however, ensures the smallest possible pressure drop and herewith the smallest energy consumption. In this respect, the construction of machine hoods is crucial in reducing the required air flow rate.

6.1 Ventilators

Ventilator is an air transporting device with external driving which capable to transport a given air volume in the unit time against a given counter-pressure. There are two basic types of ventilators:

- radial or centrifugal ventilators for higher counter-pressures, generally up to 5000 N/m^2,

E. Magoss et al., *Dust Extraction and Handling in the Wood Industry*,
SpringerBriefs in Applied Sciences and Technology,
https://doi.org/10.1007/978-3-031-08915-2_6

- axial or fan ventilators with lower pressure range, mostly for workroom ventilation.

In pneumatic transportations the system pressure drop is several thousands of N/m^2 and, therefore, centrifugal ventilators are used.

The general build-up and the main parts of a centrifugal ventilator is shown in Fig. 6.1. In the rotating impeller wheel, a centrifugal power field occurs and the air entered axially into the impeller will be forced toward the outer radii. In the rotating impeller, the air performs a combined motion which is illustrated with the velocity triangle, Fig. 6.2.

We may distinguish three different blades such as forward curved blades, radial flat blades, and backward-curved blades. The velocity triangles can be constructed to the inner and outer edges of the blade. The inner edge of the blade has always a more or less radial position and, therefore, all types of blades have the same velocity relations. Due to the rotation, the impeller outer edge has a tangential velocity

Fig. 6.1 The main parts and build-up of a centrifugal ventilator

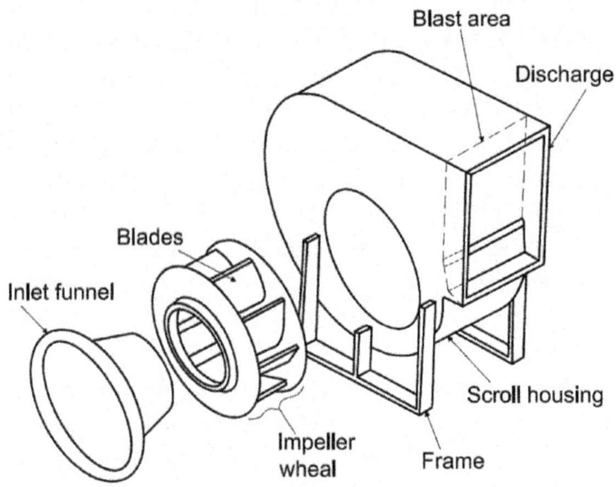

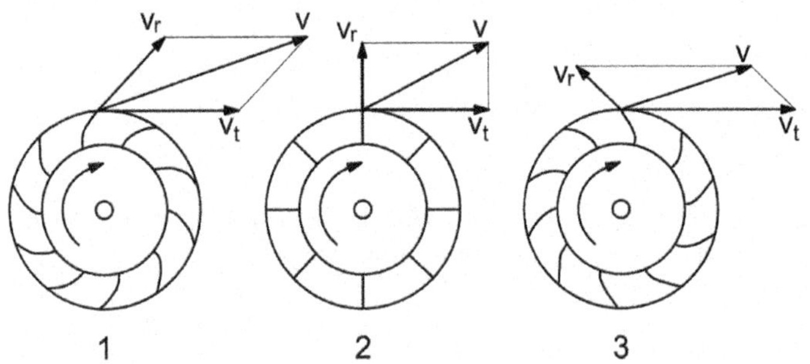

Fig. 6.2 The three blade types of centrifugal ventilators and the corresponding velocity triangles. 1—Forward curved blades, 2—radial flat blades, 3—backward-curved blades

$$v_t = \frac{D \cdot \pi \cdot n}{60} \quad \text{m/s}$$

where D is the diameter of impeller,

n is the rotation speed, rpm.

The air leaves the rotating blade tip more or less tangentially and it is the relative velocity of air v_r in relation to the blade. The absolute velocity of air is given by the vector summation of tangential and relative velocities.

As it can be seen in Fig. 6.2, the ratio of absolute velocity to the tangential velocity varies considerably for the different types of blades. The highest absolute air velocity is obtained with forward curved blades. The air velocity, according to Euler-equation, represents a dynamic pressure

$$\Delta p_d = \frac{\rho}{2} \cdot v^2 \quad \text{N/m}^2$$

which cannot be utilized for material transportation because it requires static pressure. Therefore, ventilators with backward-curved blades are mainly used.

It should be stressed that only the tangential velocity can be uniquely calculated, the relative and absolute velocities depend on the counter-pressure of the system. In the case of full throttling, there is no air transportation and the relative and absolute air velocities are zero. At the same time, the tangential velocity remains the same. The pressure rise in the ventilator must be in equilibrium with the counter-pressure of the transporting system.

The most important characteristic of a ventilator is its pressure-air volume rate relationship, Fig. 6.3. The stable operation point is possible at a volume rate where the pressure rise in the ventilator is in equilibrium with the pressure loss of the whole transport system. Readers interested in the theory of ventilators are referred to the special literature (Eck 1962; Sitkei 1997).

The selection of an appropriate ventilator is made in the knowledge of system requirements concerning the air flow rate and the expected whole pressure drop. The airflow rate (m^3/h) is generally prescribed by the manufacturer for the woodworking

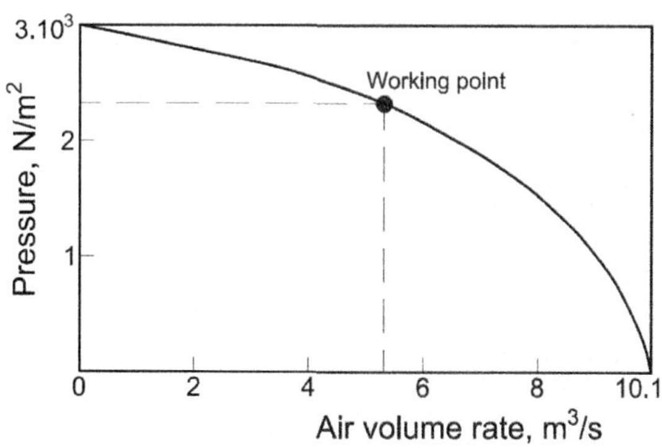

Fig. 6.3 Transport characteristics of a ventilator

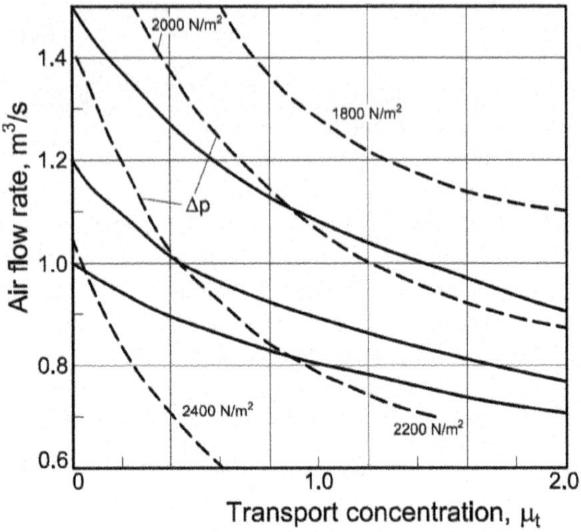

Fig. 6.4 Decrease of air volume rate as a function of transport concentration at different counter-pressures. Dotted lines indicate constant pressure drop

machines operated simultaneously. The resistance of tube system with bents and branching elements should be calculated.

In the wood industry collection and transportation of dust occurs in suction operation mode and, therefore, the dust-laden air flows through the ventilator. Transportation of dust requires energy and pressure drop and, as a consequence, the transport capacity of a ventilator varies as a function of dust concentration in the air. The main regularities of these processes can be determined only experimentally.

A ventilator is operating along its $\Delta p - V$ charactheristics depending on the counter-pressure drop and where an equilibrium is established is termed the working point. If dust is charged into the air, the working point is moving toward lower air volume rate and pressure drop. This can be seen in Fig. 6.4 (Boronkai 1985a, b) where the variation of air volume rate is given as a function of dust transport concentration at different counter-pressure drops. The dotted lines indicate constant pressure drops.

In practice, the dust mass flow rate is a more convenient input in kg/s which can also be calculated using the operational characteristics of woodworking machines. Such a relationship is seen in Fig. 6.5. As the dust mass flow increases the air volume decreases and the intersection point indicates the new equilibrium. For orientation, the constant lines of transport concentrations are also given (see Eqs. (4.16) or (4.16a)).

It is interesting to examine the $\Delta p - V$ characteristics of the ventilator as a function of dust load. Figure 6.6 shows these characteristics (full line) with different working points depending on the counter-pressure. In the case of dust-laden air, this characteristic is shifting downward depending on the dust transport concentration. For example, if the initial working point was at no. 4 with clean air, then the pressure drop is 2170 N/m^2 and the airflow rate is 1.5 m^3/s = 5400 m^3/h. If we have a dust transport concentration of 0.9, then the working point of ventilator moves to $\Delta p = 2000$ N/m^2 and $V = 1.1$ m^3/s = 3960 m^3/h. That means a reduction of 27% in the airflow rate. At the same time, the reduction in pressure drop is only 8%. If the

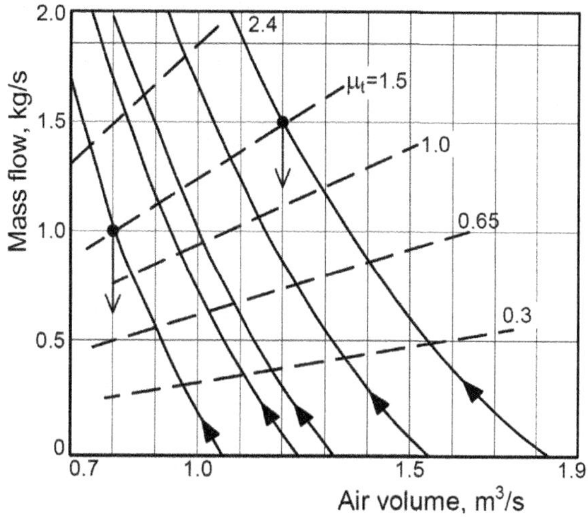

Fig. 6.5 Decrease of air volume rate in relation to mass flow rate (kg/s) with the constant lines of transport concentration

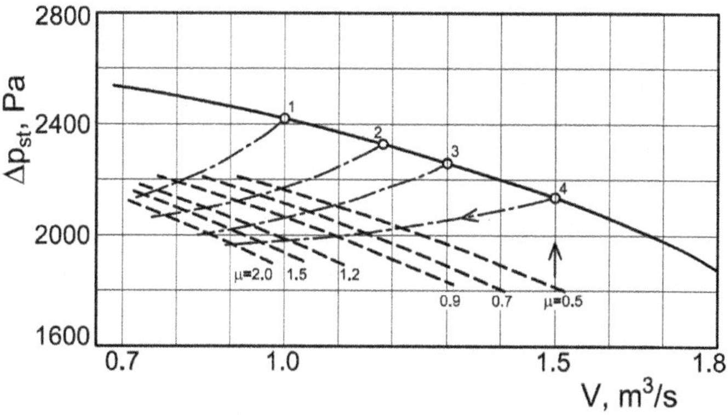

Fig. 6.6 Pressure drop–air flow rate characteristics of a transport ventilator ($D = 600$ mm, $n = 2370$ rpm) for clean and dust-laden air

counter-pressure of the system is changing, the working point is moving along the dotted line corresponding to a given dust transport concentration. The dotted lines represent the prevailing characteristics of the ventilator operating with dust-laden air (Boronkai 1993).

For the characterization of ventilators it is customary to use dimensionless numbers in the following form:

the suction number is related to the air volume rate

$$\varphi = \frac{4V}{D^2 \cdot \pi \cdot u_2} \tag{6.1}$$

the pressure number is related to the pressure drop of the system

$$\Psi = \frac{2 \cdot \Delta p}{\rho_0 \cdot (1 + \mu_t) \cdot u_2^2} \tag{6.2}$$

where D is the diameter of impeller,

u_2 is the tangential velocity of impeller.

In the case of dust-laden air, an apparent air density is used as a function of transport concentration μ_t.

For a given ventilator the values of D and u_2 are known and, using the measured values of Δp and V, the dimensionless numbers can be calculated and plotted. For a given transport ventilator such a diagram is shown in Fig. 6.7 (Boronkai 1985a, b). The ventilator had radial flat blades, $D = 0.5$ m, $n = 2370$ rpm, $u_2 = 62$ m/s. The curves can be described by the following equation

$$\Psi = \frac{0.48}{(1 + \mu_t)^{1.25} \cdot \varphi^{0.3}} \tag{6.3}$$

On the other hand, by substituting the given values of D and u_2 into Eqs. (6.1) and (6.2), we get

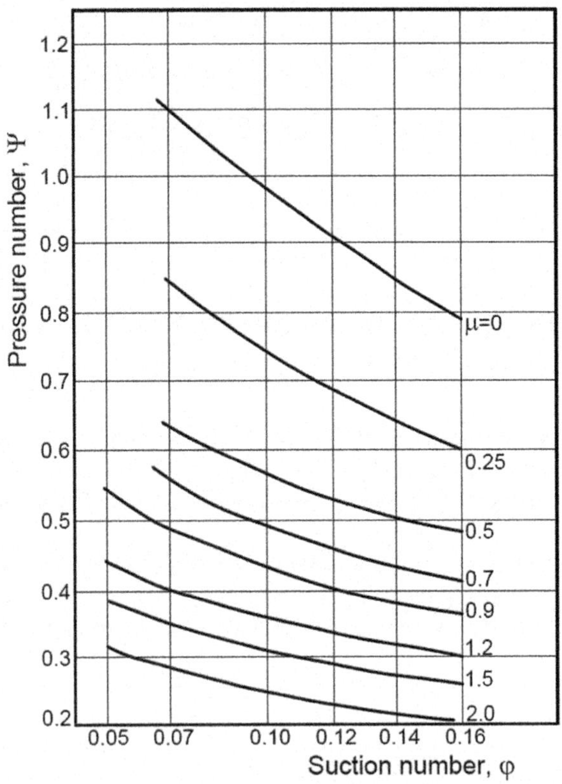

Fig. 6.7 Correlation between the pressure and suction numbers for different transport concentrations

$$\Psi = \frac{\Delta p}{2380 \cdot (1 + \mu_t)} \quad \text{and} \quad \varphi = \frac{V}{12.174}$$

Combining the above equations, yields

$$\Delta p = \frac{1143}{(1 + \mu_t)^{0.25} \cdot \varphi^{0.3}} \tag{6.4}$$

This equation well describes the measured ventilator characteristics given in Fig. 6.7. If we use the basic $\Delta p - V$ characteristics for clean air, then the initial equilibrium point moves toward smaller air volume rates if dust is transported. The new volume corresponding to the transport concentration μ_t is approximately calculated as

$$V = \frac{V_0}{\sqrt{1 + \mu_t}} \tag{6.5}$$

where V_0 corresponds to the equilibrium point for clean air. Using Eq. (6.5), the previous Eq. (6.4) has the form

$$\Delta p = \frac{2420}{(1 + \mu_t)^{0.1} \cdot V_0^{0.3}} \tag{4a}$$

This equation well describes the experimentally obtained Fig. 6.6. For example: if we start at $V_0 = 1.5$ m^3/s and $\mu_t = 0$, then the corresponding pressure drop is $\Delta p = 2143$ N/m^2. If the system operates at $\mu_t = 1.2$ then $\Delta p = 1980$ N/m^2 and the air flow rate is $V = 1.5/\sqrt{2.2} = 1.01$ m^3/s. These values are in full agreement with the measured values in Fig. 6.6.

The power required for driving is calculated from the following equation

$$P = \frac{\Delta p \cdot V}{\eta} \quad W \tag{6.6}$$

where η is the coefficient of efficiency taking all losses within the ventilator into account.

For other ventilators different constants are valid but the general form of equations remains the same. This evaluation method is therefore useful for other transport ventilators and facilitates the extrapolation for data not determined experimentally.

6.2 Feeding Devices

In dust extraction and conveying systems different devices are used to perform various feeding operations. The most special and unique feeding device is the extraction hood which is an enclosed space around the operating cutting edge and, with the help of air stream, the wood particles will be brought into the extraction tube (see in Sect. 3.1). The more common feeding devices are the rotary valve, screw conveyor, screw discharger with circular motion. Some of them perform combined duty: discharge from bins and filters, controlled flow of the material (metering device), flow aid device with discharge for highly cohesive bulk materials such as wood dust and chips (see in Sect. 2.3).

The operation of feeding devices may be influenced by the conveying system. Using vacuum system, the material feed is available at atmospheric pressure and, therefore, the conveying tube can be fed directly from the supply hopper. Obviously, a suction nozzle can also feed directly from a storage with free material surface or from a stockpile.

Using positive pressure system, the pipeline feeding may cause some problems. The material contained in a storage bin is at atmospheric pressure and, therefore, the material has to be fed against a pressure gradient. In this case, in order to compensate for losses in air supply, it is wise to increase the planned air flow rate in a certain extent.

General requirement for feeding devices is a continuous and uniform feeding of the material as possible. This is especially important in dilute phase conveying systems, where any surges in the feeding rate can block the system very quickly.

Feeding devices with positive displacements, such as rotary valves and screw conveyors, have the advantages of both metering the material into the conveying tube and ensuring airlock between storage bins and conveying system.

The rotary valve is a very commonly used feeding device. It consists of bladed rotor working in a housing. The material feed rate can easily be modified by varying the rotor speed. Figure 6.8 shows the general layout of a rotary valve and its arrangement on a feed hopper (Magoss et al. 2021).

The throughput of a rotary valve is calculated as

$$V_m = V_r \cdot n \cdot \Phi \cdot 60 \quad \text{m}^3/\text{h}$$

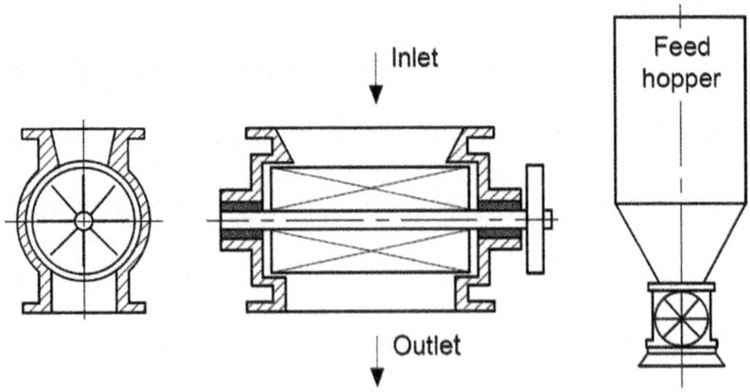

Fig. 6.8 Drop-through rotary valve and its arrangement on a feed hopper

V_r is the volumetric capacity of rotor, m^3,
n is the rotation speed of the rotor, rpm,
Φ is the filling of the rotor, decimal.

The rotary valve is a displacement device and the quantity of material fed is proportional to the rotation speed up to a given critical speed. Above a given rotation speed, there is not sufficient time to fill and discharge the pockets of the rotor completely and, therefore, the throughput decreases compared to the theoretical value. The filling of the rotor depends on the flowing properties of the material and rotation speed.

As a guide, Fig. 6.9 shows the variation of filling of a rotary valve using different sawdusts. The smaller the particle size, the lower the filling of the rotor. Due to the high variability in the flow properties, it is important to check the throughput with simple measurements. Any change in the material properties requires a change in the rotation speed of the rotor.

Screw feeders are also positive displacement devices and their feed rate can be controlled by varying the speed. They can be used for both vacuum and positive pressure systems. The screw feeders have two basic types: the simple screw feeder with constant pitch and screw feeders with variable pitch. The variable pitch ensures special capabilities: if the screw increases in pitch along its length, this feature allows an accurate metering and conveying without packing and increased resistance: if the screw decreases in pitch, the material is compressed to form a tight seal to prevent air leakage through a non-return valve feeding the material into a positive pressure conveying line.

The assembly and working principle of a screw feeder is shown in Fig. 6.10. The material fed into the feed section is carried over the shaft. At high rotor speeds, there may not be enough time to fill the feed section and less material is conveyed. That means a filling efficiency should be taken into account. By limiting the flow control

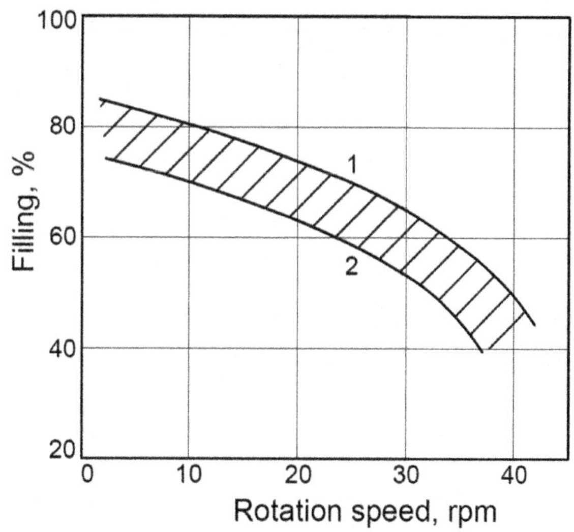

Fig. 6.9 Filling of a rotary valve as a function of rotation speed. 1—sawdust 300 kg/m^3, 2—sawdust 150 kg/m^3

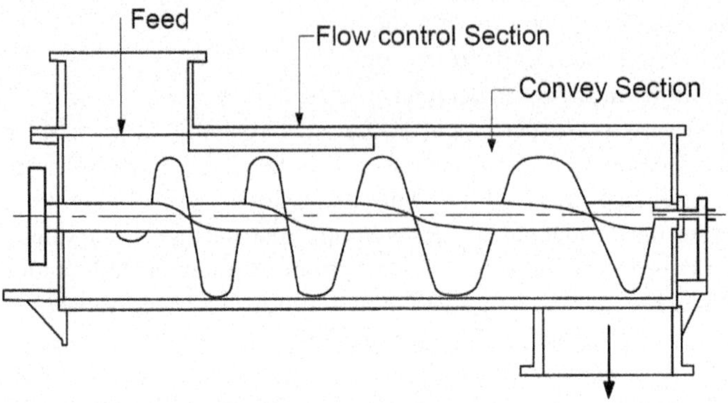

Fig. 6.10 Screw feeder assembly and working principle

section determines the amount of transported material per revolution. The conveying section has an increased pitch to prevent material compaction and increased friction.

The throughput of a srew conveyor can be calculated by the following equation

$$V_m = \frac{\pi}{4} \cdot \left(D^2 - d^2\right) \cdot h \cdot n \cdot \Phi \cdot 60 \quad \mathrm{m^3/h}$$

where D is the screw blade diameter,
d is the shaft diameter,
h is the pitch,
Φ is the filling efficiency.

The stored bulk material is highly cohesive with unfavorable flow properties and the outflow from a bin can generally be ensured by the use of flow aid devices (Sitkei 1986). Therefore, a flow aid device is incorporated into the silo or hopper design in order to provide a controlled amount of discharge from the silo.

The use of flow aid devices means additional costs for their installation but they may also be used to reduce the overall cost of storage equipment. The free outflow of the material requires a conical bottom section which, in the case of small cone angle, considerably increases the height of the silo and also its cost. The use of a flow aid device enables an almost flat bottomed silo with much lower height.

The most common flow problem is the arching (bridging) of the material above the outlet orifice. In some cases, the formation of a channel may occur through the solid mass (ratholing). The promotion of flow can be initiated by

- fluidization or aeration which involves the mixing of air to the solid phase to facilitate flow,
- air-assisted promotion of flow in which high-pressure air blasts are used to break down arching and dead areas,

- a vibration device activated pneumatically or electrically and fitted externally to the conical hopper. Another solution in the bridge breaker device which induces vibrations into the bulk material in the vicinity of bridge formation,
- hopper liners with a low friction coefficient which facilitate and promote a mass flow in the conical hopper,
- a mechanical device promoting flow that is installed into the silo designed specifically for the given case. There are different types of mechanical devices such as rotary plows, screw activators, paddle feeders, and screw dischargers with circular motion.

It should be noted that the continuous transport of dust and chips without storage in larger volumes helps to prevent flow difficulties. For example, the continuous discharge of separated dust from filter is fully possible without the use of flow aid device. Bigger storage bins, however, require flow aid devices due to the settlement and compaction of dust during its storage.

6.3 Layout of Conveying System

Dust and chips produced by woodworking operations are captured by air streams in the hoods and transported by suction into the main transport tube, and further for filtering the particles and intermediate storage in a bin. Depending on the number and location of woodworking machines, an appropriate layout of the conveying system is planned which fulfills technological and economic requirements. In many cases, the system should have some flexibility to connect or disconnect machines to the system without considerable redesign.

A relatively simple layout is shown in Fig. 6.11. The arrangement of woodworking machines allows to use one main collecting duct and the dimensioning of the system

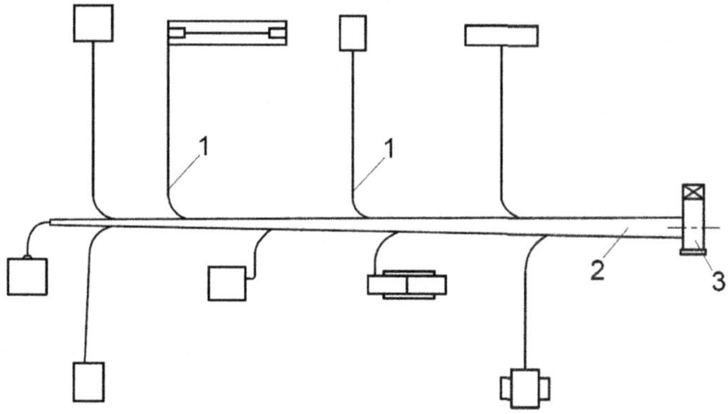

Fig. 6.11 Dust extraction and conveying system with one main duct. 1—extraction tube, 2—main, collecting duct, 3—ventilator

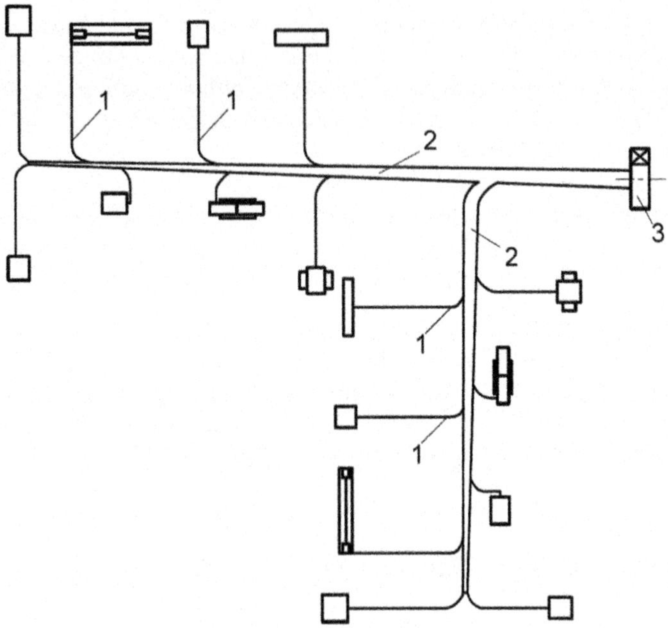

Fig. 6.12 Dust extraction and conveying system with two main collecting ducts. 1—extraction tube, 2—main duct, 3—ventilator

is easy and reliable. Obviously, the system shown here is incomplete concerning no filter and storage unit.

A more complicated arrangement of machines may require two or more main collecting ducts as shown in Fig. 6.12. The higher number of machines means always more uncertainties in the system design. That is also obvious that the ideally tapered main duct is, in practice, a stepped duct, i.e., divided into sections of constant cross-section. It is also important that in the connection points (joints) a common pressure and velocity prevail.

In practice, the extraction part of the system is always operating in negative pressure mode. An advantage is that all gas leakage is inward, so the get out of dust into the environment is practically eliminated. At the same time, under vacuum conditions, the filter might require larger size due to somewhat higher volume of air to be filtered. Therefore, the combined negative and positive pressure systems are widely used.

Figure 6.13 illustrates a combined negative and positive pressure system (suck – blow) with dual filter unit. The cyclone serves as a pre-filter for separation of bigger particles above 30 μm. The multi-cyclone used is capable to separate particles down to 10 μm. The ventilator can also be accommodated outdoors decreasing the noise level in the workroom.

The use of recirculation of air in workroom requires high efficiency of particle separation which needs fabric filter. It is advisable to use a pre-separator unit which

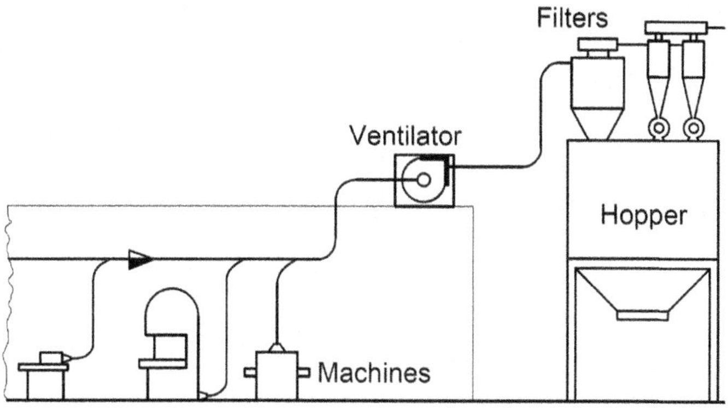

Fig. 6.13 Combined negative and positive pressure system with dual filter unit

may be a cyclone. Such an arrangement, as a combined negative and positive pressure system, is shown in Fig. 6.14. Both the filter and ventilator are located outdoor minimizing pollution and noise inside the workroom.

For smaller number of machines an interesting simplified conveying system, since the 1950s, has been developed and widely used (Tauber 1978). A common collector unit of different forms was designed which, as a single unit, has replaced the numerous joint elements of the common system. In the collector unit, the same pressure prevails for all machines which highly facilitates the calculations. The general layout of this system is given in Fig. 6.15 which also may be a combined (suck–blow) system.

Several collector types have been constructed to ensure a flexible application for different local conditions. The typical design varieties are shown in Fig. 6.16. Type ***c*** was a favorite in practical applications which ensures a smooth change in flow direction. The system has, however, limited flexibility in changing the number of

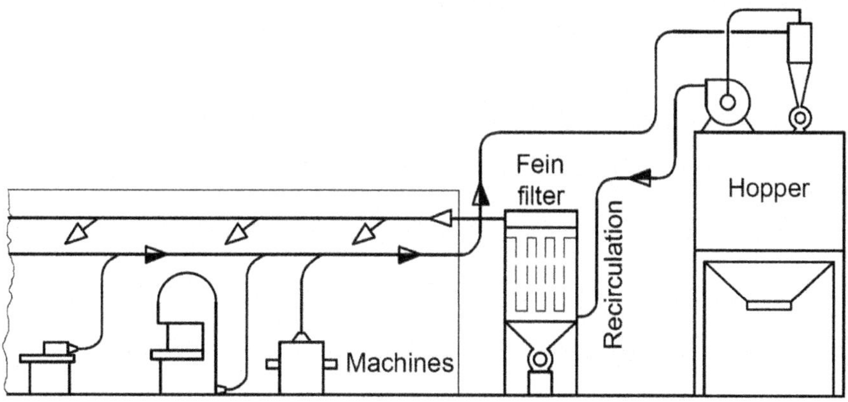

Fig. 6.14 A combined negative and positive pressure system with fein filter and recirculation

Fig. 6.15 Dust extraction and conveying system with a common collector unit

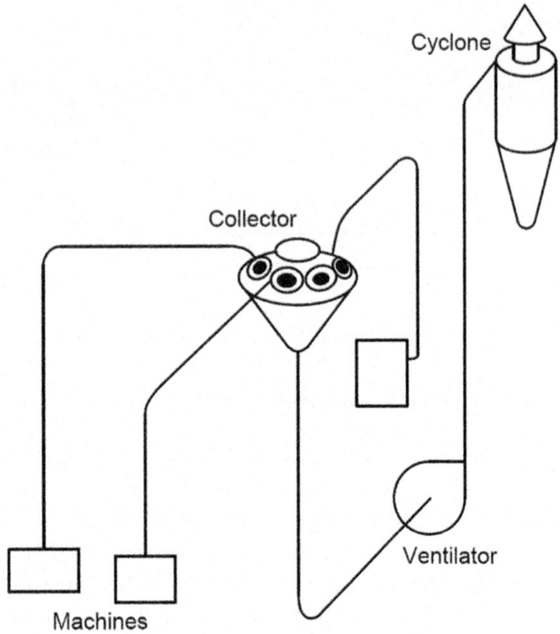

machines in operation. The design value of air flow rate may be modified with max 20%.

To cover the need for high flexibility requirements, a full automatic self-controlled system was also proposed (Larionov 1970) using several ventilators connected to the main duct. If the number of operating machines are changed, an automatic control device, based on the pressure change in the main duct, switches on or off a ventilator. For the economic use of this system no reliable data are available.

In some cases the vacuum pick-up and conveying of dust from open storage or stockpile may be of interest, using a suction nozzle. The suction nozzle is buried into the bulk material and one part of the suction air enters over the pile into the pipeline supplying the necessary air to avoid blocking. A typical vacuum pick-up and conveying system is shown in Fig. 6.17 (Sitkei 1997). The dust is separated in a cyclone and is fed through a rotary valve into the transport tube with positive pressure.

A frequent problem encountered in the practice is the need for changing the number of operating machines at a given conveying system. The air mover is generally a radial flow ventilator with flat blades (see Figs. 6.1 and 6.2). Its pressure drop – volume rate characteristics at constant speed is similarly given in Fig. 6.3. That means that the airflow rate is very dependent on the conveying line pressure drop. As a consequence, the connection of further machines, or disconnection, considerably influences the air volume rate and pressure drop, and herewith the flow velocity in the transport tubes. A decrease in flow velocity will contribute to the risk of blocking the pipeline. In such cases, it is advisable to keep some reserve in the minimum air

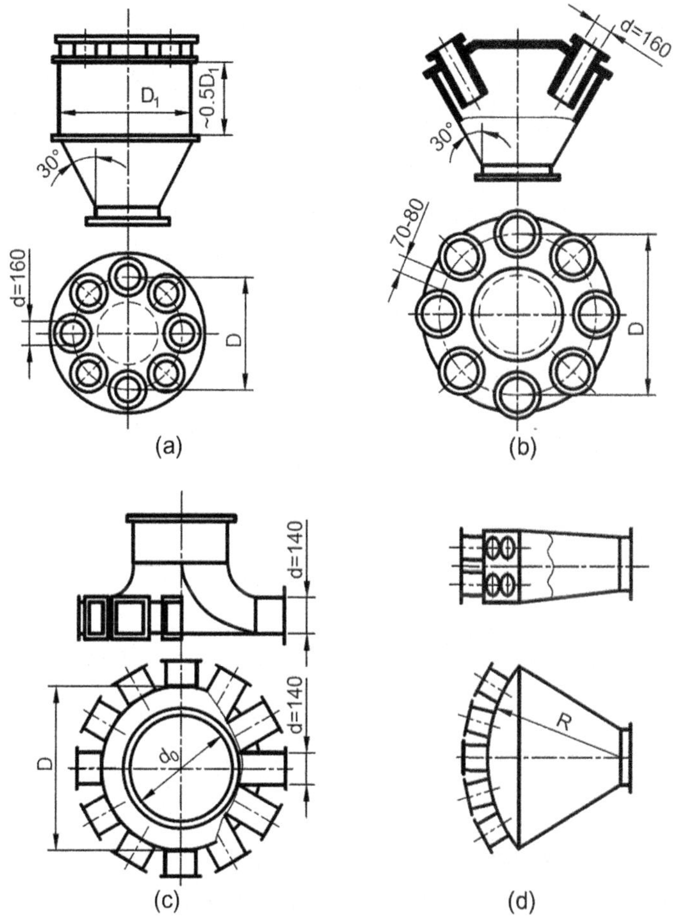

Fig. 6.16 Different types of collectors. **a** vertical cylindrical collector, **b** vertical conical collector, **c** horizontal spherical collector, **d** horizontal fan-shape collector

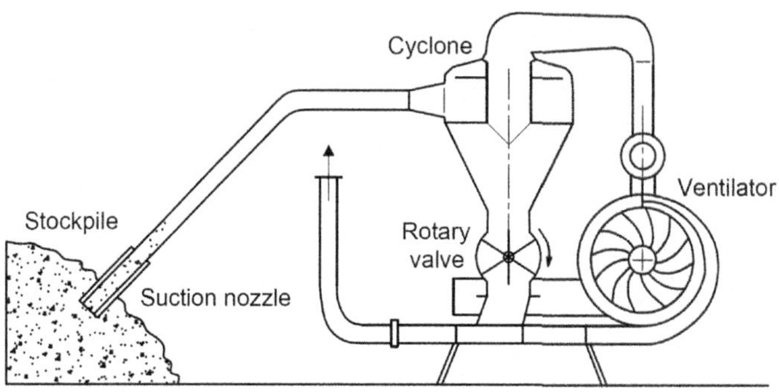

Fig. 6.17 Vacuum pick-up and conveying from stockpile

velocity to balance the effect of connected new machines. In the following, as an example, we use the measured characteristics of a transport ventilator depicted in Fig. 6.6. We select system with nominal design data such that it corresponds to the working point No. 4 on the full line characteristics for clean air. That means that the clean air flow rate is 1.5 m^3/s and the pressure drop is 2150 N/m^2. The arrangement of our examined system is given in Fig. 6.18. The clean air characteristics are well described for the feasible range with the following empirical equation:

$$\Delta p = 2.63 \cdot 10^3 \cdot \sqrt{1 - (V/2.6)^2} \ \text{N/m}^2 \tag{6.7}$$

where V is the air flow rate, m^3/s.

In order to calculate pressure losses in the conveying system, we use Eqs. (4.5) and (4.6). For the sake of simplicity, we assume that flow velocities in all pipelines are the same, and the extraction pipelines have the same length and diameter. With these assumptions, we may write

$$\Delta p = \left(\frac{\lambda}{d_1} \cdot l_1 + \xi_1 + \frac{\lambda}{d_2} \cdot l_2 + \xi_2 \right) \frac{\rho_0}{2} \cdot v^2 \ \text{N/m}^2$$

We use the following input data: in each of the 5 extraction tubes $V = 0.3$ m^3/s, the total flow rate is 1.5 m^3/s, the tube length are $l_1 = 10$ m, $l_2 = 15$ m, the flow velocity $v = 24$ m/s, the tube diameters are $d_1 = 0.1262$ m, $d_2 = 0.282$ m, resistance coefficient of hoods and the collector are $\xi_1 = 3.1$ and, $\xi_2 = 0.6$, and using Fig. 4.10, $\lambda/d_1 = 0.157$ and $\lambda/d_2 = 0.055$.

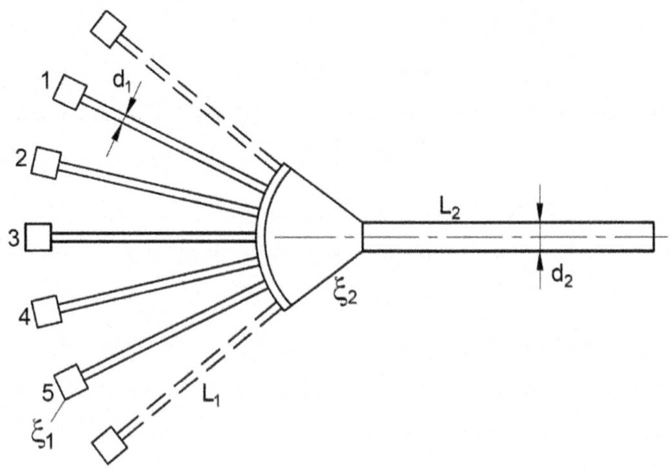

Fig. 6.18 System layout for worked example, with five constantly operating machines and two reserve connections

Using these data in the above equation, the total pressure drop in the pipeline is 2142 N/m^2 which equals the working point of ventilator.

Now we connect a new similar machine to the conveying system and the airflow rate increases to a given value determined by the equilibrium conditions. A new equilibrium is given between ventilator and pipeline at $V_0 = 1.664$ m^3/s, $\Delta p = 2021$ N/m^2 with $v_1 = 22.2$ m/s and $v_2 = 26.64$ m/s (main duct). It is important that at the machines the air velocity is decreased. Similarly, if we disconnect one machine from the five, then in equilibrium the air flow rate is $V_0 = 1.555$ m^3/s, the pressure drop is $\Delta p = 2110$ N/m^2 and the air velocities are $v_1 = 24.9$ m/s and $v_2 = 19.92$ m/s. In this case, the air velocity considerably decreased in the main duct.

The net power requirement in the nominal case is $1.5 \times 2142 = 3.213$ kW. Connecting an additional machine, it increases to 3.363 kW, while disconnecting one machine the system requires 3.28 kW. That means that the power requirements are almost the same.

The derived Eqs. (4a) and (6.5) are suitable to follow similar processes in the case of material transport. Due to material transport, both the pressure drop and air flow rate vary, and also the flow velocities too. Note that in cases deviating from the nominal, the velocities in the extraction and main tubes are different. Therefore, we always examine the further decrease of smaller velocities due to material transport.

The starting point is always the equilibrium flow rate V_0 for clean air transportation which decreases according to Eq. (6.5) as a function of transport concentration. Obviously, in the same ratio, it decreases the flow velocity in the given tube section. The results of the examinations are illustrated in Fig. 6.19. If we establish a critical

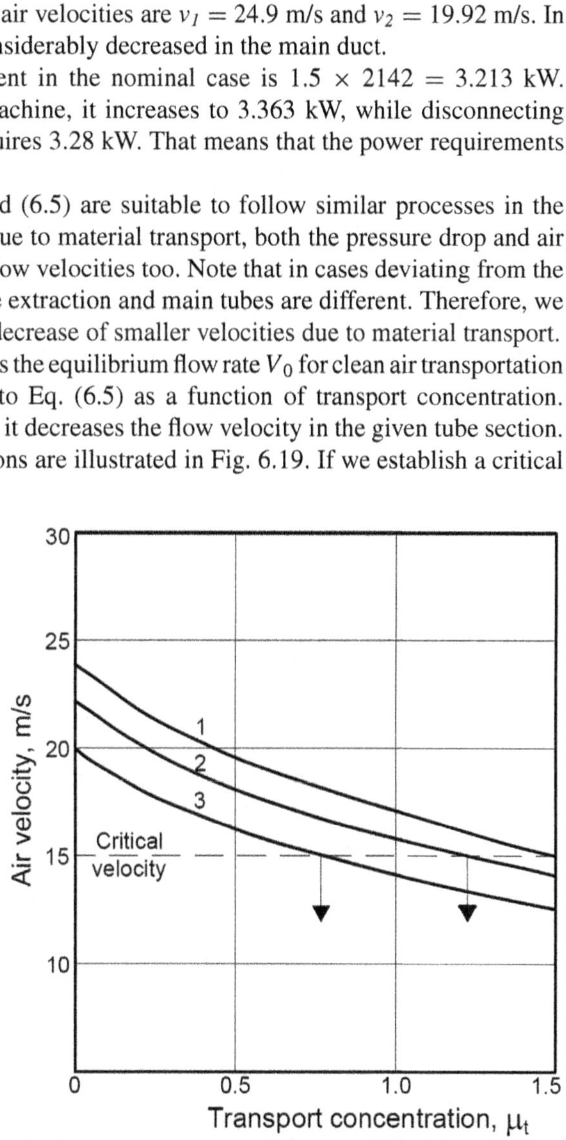

Fig. 6.19 Air velocity as a function of transport concentration. 1—constant velocity along the pipeline, $v_0 = 24$ m/s. 2—one machine additionally connected, $v_{10} = 22.2$ m/s. 3—one machine disconnected, $v_{20} =$ 19.92 m/s

flow velocity (15 m/s), we obtain allowable concentration without the risk of blocking the system. It is interesting that the connection of a new machine causes less limitation that the disconnection.

From the above example, it follows that the careful design of any extraction and conveying system is very important. Especially, the accurate knowledge of the characteristics of the selected transport ventilator is indispensable. Some reserve in the selected minimum air velocity helps to overcome occasional difficulties caused by deviations from the nominal operational parameters (deviation in material properties, overloading, etc.).

As already mentioned before, the main task of an extraction and conveying system is to remove the dust and chips produced by the woodworking machines and, as far as possible, to prevent the getting out of fine dust particles into the workroom.

The main input data for system design are the followings:

- number and location of woodworking machines in the workroom,
- the required air volume rate (m^3/s) for each machine,
- the approximate maximum chip production (kg/s) for each machine,
- the expected physical properties of the produced chips such as particle size distribution and moisture content,
- requirement for dust separation efficiency determines the selection of filter equipments,
- required storage capacity for the separated dust depending on the subsequent handling.

The first step is an appropriate mark out of pipeline arrangement within and outside the workroom with respect to machine location. This step also determines the type of pipeline system (see Figs. 6.11, 6.12, and 6.15). The required air volume rate for each machine determines the tube diameters for each section with the necessary bends and joining elements. Furthermore, a safe transport air velocity should be selected with respect to possible changes in the number of operating machines.

The next important question is the expected pressure drop along the pipeline. The main sources of pressure drop are the machine hoods, bends, joinings, straight pipeline sections, and separation equipment's. Furthermore, the bigger part of pressure drop is generally caused by the clean air flow, but the additional pressure drop due to material transportation must also be taken into account.

Concerning clean airflow, the sum of particular pressure drops is given by the following general equation:

$$\Sigma \Delta p_0 = \sum \left(\lambda \cdot \frac{L_i}{d_i} + \xi_i \right) \cdot \frac{\rho_0}{2} \cdot v_i^2$$

where ξ_i—are the resistance coefficients for hoods, bends, joining elements, and filters.

The additional pressure drop due to material transportation can be calculated, as a quick check method, with the Gasterstädt equation with $k_0 = 1$.

Using the relationships given in Sect. 4.4, a more detailed calculation can be carried out. Note, however, that the pressure drop component due to material transportation can only be determined with an error bound. The particular elements of the system show different behavior with dust-laden air. Please remember, for example, that a cyclone has less resistance with dust-laden air. Other components of the system, at the same time, response with higher resistance. Therefore, a completed new system should be checked with measurements of flow velocity and pressure drop in several characteristic points of the system.

References

Boronkai, L.: A faiparban alkalmazott por-forgács elszívó és szállítórendszerek méretezésének néhány kérdése (Pneumatic conveying systems in the wood industry) PhD Dissertation, Sopron, pp. 39–85 (1993)

Boronkai, L.: Druckverlustwerte für das pneumatische Fördersystem in der Holzindustrie. Acta Facultatis Ligniensis. Sopron, pp. 17–29 (1985a/1)

Boronkai, L.: Faipari por-forgács elszívó és szállítórendszerek transzport ventilátorai (transport ventilators in the wood industry). Third Conference on pneumatic conveying, Pécs, pp. 39–62 (1985b)

Eck, B.: Ventilatoren. Springer Verlag (1962)

Magoss, E., Sitkei, Gy., Kocsis, Z.: Collecting and handling of dust and chips in the wood industry: Progress Report No.9, Published in 2021 by the Department of Wood Engineering, University of Sopron, Hungary, p. 66 (2021)

Sitkei, G.: Mechanics of agricultural materials. Elsevier, pp. 284–402 (1986)

Sitkei, Gy.: Gyakorlati áramlástan (Applied fluid mechanics). Budapest (1997)

Ларионов, В., Проектирование и расчет самонастраивающихся систем пневмотранспорта, (Design and calculation of self-adjusting pneumatic conveying system). Деревообрабатывающая Пром., №. 11, pp. 18–20 (1970)

Таубер, Б.: Внутризаводский транспорт. (Internal transport in wood industry), Машгиз (1978)

Chapter 7
Health and Safety

The wood dust, the tiny wood particles can cause different adverse effects on human health. In the wood industry, the wood dust is often contaminated with different chemicals (glue, substances of surface treatment) which causes also health risks.

The tiny wood particles can become airborne, can penetrate into the respiratory system of the workers, and under special conditions could become explosive material. Therefore, the wood dust emission must be minimized in the workshop. The technology of the wood processing is seldom optimized on the health and fire hazards, therefore the wood dust emission must be under control through licensing as well as by the use extraction systems.

In order to understand the dust explosion process better, results of new processing of explosion research as a function of main influencing variables are presented. They may help to select operational parameters which lower the risk and destructive effect of dust explosions.

7.1 Health Hazards of Wood Dust

Wood dust is known to be a human carcinogen, based on sufficient evidence of carcinogenicity from studies in humans. It has been demonstrated through human epidemiologic studies that exposure to wood dust increases the occurrence of cancer of the nose (nasal cavities and paranasal sinuses) (NTP Board 2000).

Exposure to dust from softwood, hardwood, and wooden boards is common among European workers, but the exposure levels to hardwood dust are usually below 5 mg/m^3. However, over 0.5 million workers may be exposed to a dust level (any type of wood dust) exceeding 5 mg/m^3. High exposures occur particularly in the construction sector and furniture industry (Kauppinen et al. 2006).

Wood dust and biohazards associated with wood dust are potential health hazards in the wood processing industry. However, using proper dust extraction systems,

E. Magoss et al., *Dust Extraction and Handling in the Wood Industry*,
SpringerBriefs in Applied Sciences and Technology,
https://doi.org/10.1007/978-3-031-08915-2_7

occupational exposure to wood dust and the biohazards associated with wood dust can be prevented or minimized. When the exposure cannot be fully controlled, workers should be provided with appropriate personal protective equipment. Workers should also be educated on the potential health effects of wood dust exposure (Mandryk et al. 1999).

Particle Size Conventions concerning the penetration rate are the followings (ACGIH 1984):

- Inhalable fraction <100 μm AED (Aerodynamic Equivalent Diameter). Can be breathed into nose or mouth.
- Thoracic fraction <25 μm AED. Can penetrate head airways and enter lung airways.
- Respirable fraction <10 μm AED. Can penetrate beyond terminal bronchioles to gas exchange region.

The 2004/37/EC (1999/38/EC) EU Directive gives a limit value of 5 mg/m^3 (inhalable fraction) for occupational exposure to hardwood dust in the country of European Union. The national regulation could contain lower value, for example in Hungary this limit value is 3 mg/m^3 according to the 25/2000 (IX. 30). EüM-SzCsM Hungarian regulation.

In the industry, dust samplings are used to measure inhalable dust or/and respirable dust amount in the workshops. Personal air sampling pumps to be used with either head should be capable of maintaining a smooth flow of 2 ± 0.1 L/min over the entire sampling period. If the measured amount exceeds the limit, personal (respiratory) protection equipment must be used. It may be a face piece (dust mask), helmet, hood, suit, or a mouthpiece respirator with nose clamp.

In the wood industry, the facepiece is usually used, therefore, the dust masks detailed description is given. Dust musk as mechanical filter respirators retains particulate matter when contaminated air is passed through the filter material. The inventors such as Haslett and Tindall used wool as filter material. Today's filter materials are: wool, plastic, glass, cellulose, and combinations of two or more of these materials.

Mechanical filters remove contaminants from air in the following ways (W&M 2020):

- by dust particles that are following a line of flow in the airstream coming within one radius of a fiber and adhering to it, called *interception*;
- by larger particles unable to follow the airstream being forced to embed in one of the fibers directly, called *impaction*; this increases with diminishing fiber separation and higher air flow velocity;
- by an enhancing mechanism called *diffusion*, which is a result of the collision with gas molecules by the smallest particles, especially those below 100 nm in diameter, which are thereby impeded and delayed in their path through the filter; this effect is similar to Brownian motion and increases the probability that particles will be stopped by either of the two mechanisms above; it becomes dominant at lower airflow velocities;

- by using certain resins, waxes, and plastics as coatings on the filter material to attract particles with an electrostatic charge that holds them on the surface of the filter material;
- by using gravity and allowing particles to settle into the filter material (this effect is typically negligible); and
- by using the particles themselves, after the filter has been used, to act as a filter medium for other particles.

The marking systems of the dust mask are different on the different continents and some cases in different countries. For example, in the USA the labeling is the following (OSHA):

- There are nine classes of filters (three levels of filter efficiency, each with three categories of resistance to filter efficiency degradation).
- Levels of filter efficiency are 95, 99, and 99.97%.
- Categories of resistance to filter efficiency degradation are labeled N, R, and P.
- Use of the filter will be clearly marked on the filter, filter package, or respirator box (e.g., N95 means N-series filter at least 95% efficient).

In the European Union the European standard EN 149 defines the following classes of "filtering half masks":

Class	Filter penetration limit (at 95 L/min airflow)	Inward leakage (%)
FFP1	Filters at least 80% of airborne particles	<22
FFP2	Filters at least 94% of airborne particles	<8
FFP3	Filters at least 99% of airborne particles	<2

European standards test filter penetration with both dry sodium chloride and paraffin oil aerosols, after storing the filters at 70 and −30 °C for 24 h each. The standards also include tests on mechanical strength, breathing resistance, and clogging. EN 149 also tests the inward leakage between the mask and face (ten human subjects perform five exercises each and, for eight of these individuals, the average measured inward leakage listed above must not be exceeded).

Particle size distributions are commonly measured using a series of screens and processed by mass. All calculations in dust separation are represented also by mass (see Figs. 5.7, and 5.8, 5.10). Standards and regulations use mass units (mg/m^3) for exposure limit values.

The true mechanism of the harmful effect of dust particles getting into the lung airways and gas exchange region is probably not known in all details and it is possible that also the number of particles plays a role in the critical exposure. At the same time, processing of size distribution by number is not common. The distribution by mass can, however, be transformed into a distribution by number (Bölöni 1973). The median diameters are related by the following equation

$$\mathrm{Mi}_m/\mathrm{Mi}_n = e^{3\sigma^2} \tag{7.1}$$

where Mi_m is the median by mass,
Mi_n is the median by number
the standard deviation σ remains unchanged.

Figure 7.1 shows the same distribution by number and mass (curve No. 2 in Fig. 1.4). Taking $d = 10\ \mu m$ particle diameter, its cumulated value by number is 13%, while the same by mass is only 0.02%. The median ratio depends strong on the standard deviation.

Taking dust samples produced by sanding, the portion of fein particles by number is much higher. Depending on the sanding grade, the cumulated value by number may be as high as 65–85% concerning particle diameters of 10 μm and below. Wood species may modify the above figures in a certain extent. These facts may lead to the conclusion that in the future the particle size distribution in the recirculated air passed the filter is also of definite importance (see in Sect. 5.3).

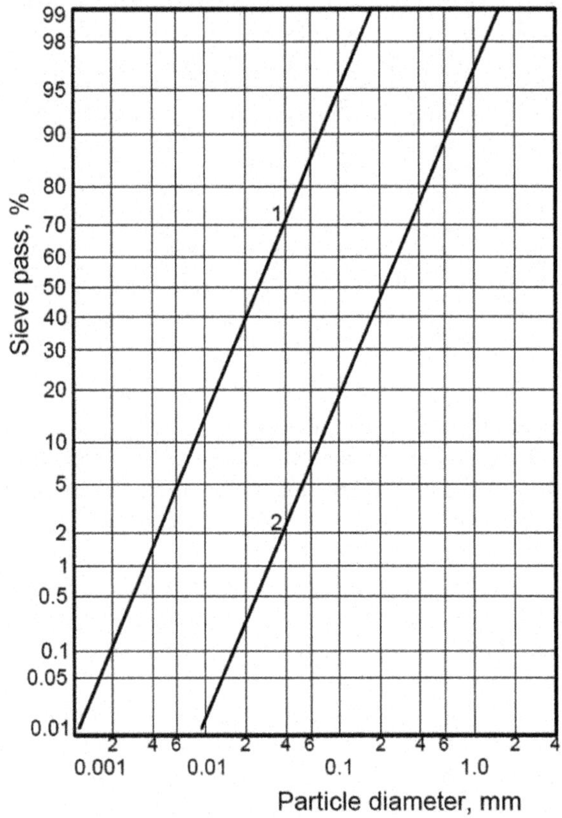

Fig. 7.1 Particle size distribution (sieve pass) made by a circular saw related to number (1) and mass of particles (2). $Mi_n = 0.0263$ mm, $Mi_m = 0.23$ mm, $\sigma = 0.85$

7.2 Fire and Explosion Protection

Wood is well known as combustable material, mankind using it as firewood since ancient times. This useful material nature covers huge risks during the wood processing. Namely, the wood dust is combustible and, according to the NFPA 654-2020 Standard, the definition of the combustible dust is the following: "Any finely divided solid material that size is 420 μm or smaller in diameter and presents a fire or explosion hazard when dispersed and ignited in air."

The technology of the wood processing seldom optimizes on the fire and explosion hazards, therefore the fire and explosion risk must be under control through licensing as well as apply fire and explosion safety equipment's.

A dust explosion is the rapid combustion of fine dust particles hovering in the air in a closed space. The wood dust explosions usually are violent, occur quickly, and can be very destructive. In order for an explosion to occur, the following 5 conditions must be met:

- The presence of wood dust;
- Ignition source;
- Oxygen;
- Fine dispersion of wood dust;
- Closed space.

An explosion is a result of ignition or self-ignition of a flammable mixture containing fuel and oxygen. Both ignition and self-ignition require heat to generate oxidation process in a small space which is self-sustained. That means that the heat generation of chemical reactions exceeds the heat loss into the environment. If it is not the case, ignition is impossible and flame quenching occurs. From this, it follows that ignition in a flowing medium requires more initial heat to be self-sustained due to the increased heat loss.

The heat source may be a spark or a hot surface. Generally, a spark is a result of electric discharge and its energy depends on the electric storage capacity and voltage drop

$$E_{SP} = \frac{1}{2}C\left(V_2^2 - V_1^2\right) \quad \mathrm{J}$$

where C is the electric capacity, Farad,

V_2, V_1 are the voltage of capacitor before and after discharge.

In practice, large metal surfaces may serve as a capacitor holding electric charge, which can discharge as a spark. Therefore, bigger metal surfaces acting as an electric capacity should be ground-connected. If a smaller stone gets into the pneumatic conveying system, its impact with the metal wall may also generate a spark, generally, of low energy.

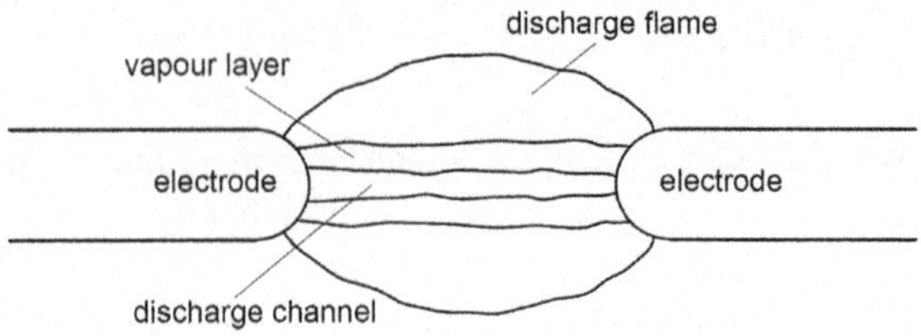

Fig. 7.2 Electric discharge (spark) between electrodes

The build-up of an electric discharge between electrodes is shown in Fig. 7.2. In the center of electrodes, a conducting channel develops with very high temperature (8000–10,000 °C). It is covered by a thin layer of vapor mostly containing the molecules of electrodes, and outside by the discharge flame containing the ionized molecules of air. The discharge of a capacitor is characterized by high electric current and short time of discharge which are the reason for the high discharge temperature.

In spark ignition, the heat effect of discharge plays the most important role, but some active molecules due to ionization processes associate in the starting of chemical reactions. There is a minimum ignition energy that is necessary to produce a self-sustained ignition front capable to propagate through the whole mixture. The minimum ignition energy depends on many factors:

- the chemical structure of fuel, especially the portion of volatile part,
- the particle size and its distribution, the surface per volume ratio of particles,
- the concentration of fuel in the air,
- the temperature of mixture,
- the pressure of mixture,
- the flow velocity of mixture.

The particle size, especially the irregular shape and hairy surface, plays an important role in the necessary minimum ignition energy. The ignition occurs in gas phase which is the mixture of volatile substances and air. A high surface per volume ratio facilitates the rapid heating-up of particles and promotes the release of volatile substances to the surface and the mixing with the air.

The concentration of fein particles in the air is also an important factor. Very low concentrations cannot be ignited at all. For particles originated from sanding the minimum concentration is around 30 g/m^3 of air. For mixed particle sizes, this limit is taken as 60 g/m^3 which is generally accepted in standards. There is also an upper explosion limit which is between 2 and 5 kg/kg depending on the specific surface area of particles which is inversely proportional to the particle diameter and density.

The ignition energy also depends on the concentration and its minimum value is obtained at moderately rich mixtures. This minimum ignition energy in resting air is accepted with a value of 60 mJ. For comparison, a propane gas-air mixture requires about 30 mJ at atmospheric pressure.

In a closed space, after ignition, the temperature and pressure rapidly increase which lower the ignition energy and accelerate the flame propagation in the space.

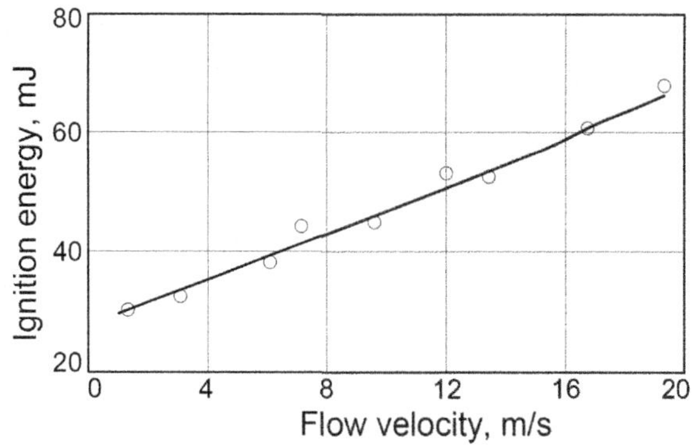

Fig. 7.3 Minimum ignition energy of propane in flowing air stream at atmospheric pressure

The airflow increases the heat remove from the vicinity of spark discharge and, therefore, a higher minimum ignition energy is needed. Figure 7.3 shows a relationship for propane–air mixture at atmospheric pressure (Sitkei 1969). At a flow velocity of 20 m/s the minimum ignition energy is at least doubled compared to the resting air. There are practical observations that dust explosion in pneumatic conveying systems is almost never encountered. At the same time, a developed fire can spread over the extraction duct. Within filters and silos the flow velocity is small and they are more prone to fire explosion. As a consequence, filter equipments and silos are generally located outdoors.

In open and resting space, after ignition, a laminar flame front is developed and it propagates with a relatively low velocity. In closed space, due to the temperature and pressure rise, a pressure wave develops turbulizing the space before the flame front and the flame becomes turbulent with much higher velocity.

The expected maximum pressure in closed space can be estimated in the following manner. The heat value of wood is, as an average, 18 MJ/kg. In the rapid combustion, only the volatile part of the fuel takes part which is some 80% of the whole mass. Therefore the effective heat value is reduced to 14.5 MJ/kg. The stoichiometric ratio is appr. 4.6 m^3 air/kg dry matter or 5.5 kg/kg. This value corresponds to 180 g fuel/kg air. That means that higher concentrations will only burn partially.

At constant volume, the pressure rise due to burning is mainly determined by the temperature rise in the space expressed in °K. Variations in specific heat have a less effect. The results of calculations are depicted in Fig. 7.4. Practical observations show maximum pressure rise up to 9 bar in agreement with these calculations. The maximum pressure rise velocity ($\mathrm{d}p/\mathrm{d}t$) may be as high as 400–500 bar/s which corresponds to a visible flame propagation velocity in the range of 9–12 m/s.

Standards relate the maximum rate of pressure rise to the length of flame propagation. Ignition more likely occurs near a wall holding electric charge. A simple assumption is for a cubical shape that this length is the cube root of the volume and the characteristic number reads

$$K_{\mathrm{St}} = V^{1/3} \cdot \left(\frac{\mathrm{d}p}{\mathrm{d}t}\right) \quad \text{bar m/s}$$

Fig. 7.4 Expected explosion pressure as a function of dust concentration in closed space

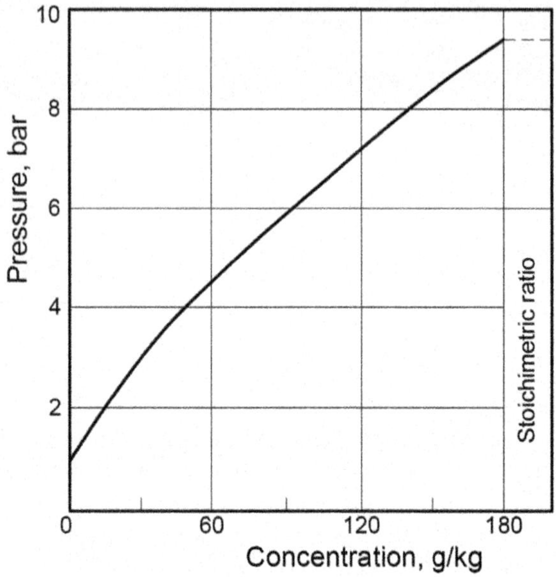

The time of burning may conditionally be divided into three sub-sections. The first sub-section contains the ignition delay and the development of flame to a stable flame front, during the second sub-section the burning occurs nearly with maximum mass burning velocity giving the maximum rate of pressure rise d*p*/d*t*, and in the third, a post-burning occurs with scattered partial flame fronts. In the last sub-section, the mass burning velocity decreases and tends to zero. The portion of these sub-sections is appr. 35, 40, and 25% related to the whole burning time.

The dynamics of dust explosion can be measured in an explosion chamber fitted with the necessary measuring units. The modeling of real cases is not a simple task, especially the mixture preparation and ignition (Callé et al. 2005). A spherical chamber with a volume of 20 L (33.7 cm dia.) was used and the given dust quantity is charged into the chamber with pressurized air of 20 bars. The ignition of mixture occurs in the center of explosion chamber with high energy release of 10 kJ. The intensive turbulent motion of air within the chamber helps to maintain higher dust concentrations in floating conditions. After ignition, a spherical flame front propagates through the chamber.

The dynamics of combustion depend on the mass velocity of burning which varies in time. Rapid combustion is possible only in gaseous phase and for this the release of volatile components is necessary. The fein particles with irregular shape, hairy surfaces, and internal voids release the volatile components easy and quickly, depending on their size and concentration. Figure 7.5 shows typical explosion pressure curves as a function of time for three narrow dust fractions and a dust concentration of 580 g/kg (Callé et al. 2005). In these experiments, this concentration gave the fastest combustion. Note that the stoichiometric concentration is around 200 g/kg. The highest pressure rise (d*p*/d*t*) occurs when the flame front is

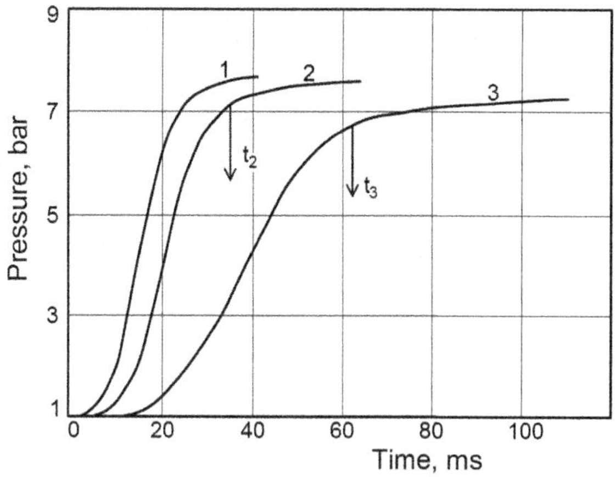

Fig. 7.5 Influence of dust fraction on the dynamics of explosion at optimum dust concentration (580 g/kg). 1—25–45 μm, 1500 m^2/kg. 2—45–71 μm, 910 m^2/kg. 3—90–125 μm, 490 m^2/kg

big enough to produce large mass burning velocity and intensive heat release. The approximate specific surface area of fractions is also given. It is interesting to note that the maximum rate of pressure rise d*p*/d*t* is directly proportional to the specific surface area of fractions with good approximation.

A deeper analysis revealed that the specific surface area (m^2/kg) of a given dust distribution can simply and exactly be calculated by the following equation:

$$A = \frac{6}{d_x \rho_m} = \frac{6 \cdot e^{\sigma^2/2}}{\mathrm{Mi}_m \cdot \rho_m} \quad \mathrm{m^2/kg} \tag{7.2}$$

where Mi_m is the median by mass, m,

σ is the standard deviation,

ρ_m is the density of particles which is, for 100 μm and below, only around 100–120 kg/m^3 (see in Sect. 1.6).

d_x means a characteristic diameter and d_x is uniquely determined by Mi_m and σ. The median of the surface area distribution Mi_A is bound to the median by mass Mi_m in the following manner

$$\mathrm{Mi}_A = \mathrm{Mi}_m / e^{\sigma^2}$$

or to the median by number Mi_n

$$\mathrm{Mi}_A = \mathrm{Mi}_n \cdot e^{2\sigma^2}$$

Keeping in mind Eq. (1.4), it follows that $\mathrm{Mi_A} = \overline{X}$. It means that the most probable particle size ($\overline{X}$) divides the cumulated surface area into two equal parts. For quick estimate, Fig. 7.6 shows the major effect of the median by mass and standard deviation on the expected specific surface area of dust distributions. An increase in

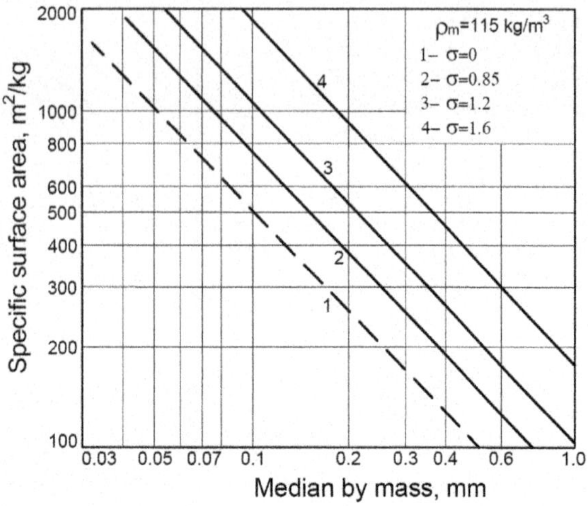

Fig. 7.6 Effect of the median by mass and standard deviation on the specific surface area of dust distributions

standard deviation highly increases the portion of small particles by number which rapidly enlarges the specific surface area, even at the same median by mass. A mixed dust concentration, when also sanding machines are used, may have a wide particle distribution which is favorable for dust explosion.

It is interesting that rich mixtures are required for high burning intensity. This indicates the importance of the presence of volatile components release of which depends on particle size and concentration. Below the stoichiometric concentration, the maximum pressure decreases and the burning time increases. Therefore, the approximate knowledge of dust characteristics in the system is important.

Although the experimental results are limited, it was succeeded to find an approximate relationship between particle size and dust concentration for constant burning times, Fig. 7.7. The burning time is characteristic for the dynamics of explosion and it determines also an average visible flame velocity. The visible flame velocity consists of two components: one is the transport velocity caused by the expansion of burned gases, and the other is the true flame velocity. The resultant flame velocity has its maximum around the halfway of possible propagation length with the highest burning rate, similarly to spark ignition engines (Sitkei 1969). From Fig. 7.7 it follows that a rapid and high-intensity explosion occurs in rich mixtures with dominant fein particles in the 10–30 μm range.

A further evaluation of experimental results revealed that, within the error bound of experimental results, the maximum rate of pressure rise depends only on the burning time. This relationship is depicted in Fig. 7.8. Due to the ignition delay and an additional time necessary for flame front development, a time-shift should be taken into account and the relationship is given by the following exponential equation

$$\left(\frac{\mathrm{d}p}{\mathrm{d}t}\right)_{\max} = 600 \cdot e^{-(t-15)/T} \quad \text{with} \quad T = 37\,\text{ms}$$

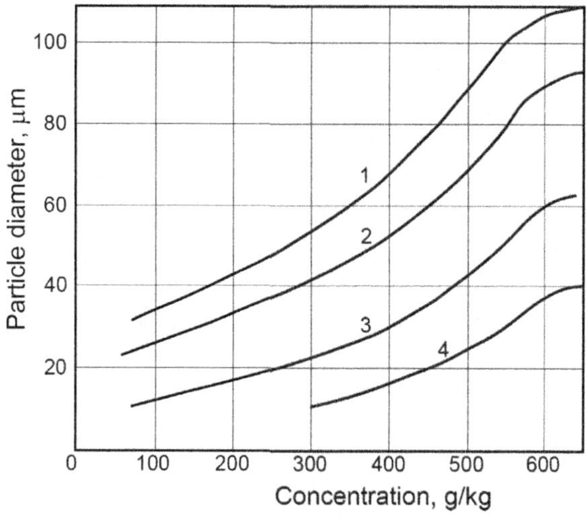

Fig. 7.7 Relationship between particle diameter and dust concentration for constant burning times and the corresponding average visible flame velocities. 1—$t = 60$ ms, $v_{fl} = 2.75$ m/s. 2—$t = 50$ ms, $v_{fl} = 3.3$ m/s. 3—$t = 35$ ms, $v_{fl} = 4.7$ m/s. 4—$t = 25$ ms, $v_{fl} = 6.6$ m/s

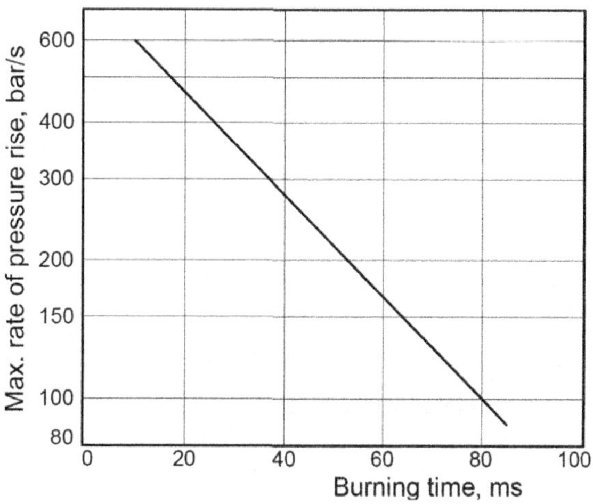

Fig. 7.8 Maximum rate of pressure rise as a function of burning time

and the burning time, based on Fig. 7.7, is calculated as

$$t = 7.75 \cdot d^{0.65} \cdot e^{-C/C_0} \quad \text{with} \quad C_0 = 528\,\text{g/kg} \text{ and } 40 \leq C \leq 600$$

where the dominant average particle diameter d must be substituted in micron.

Note that the above figures and the equation are based on model measurements, but the general regularities are valid also for practical cases.

In some cases, a hot surface may be the source of ignition. The hot surface generates chemical pre-reactions in a given volume which will be self-ignited and followed by a higher rate of pressure rise it is important to add that a bigger hot surface needs lower temperature for self-ignition. The minimum ignition temperature for dust cloud is 400 °C, but a 5 mm thick dust layer can be ignited with 250 °C. The airflow velocity also here increases the minimum ignition temperature. If a bigger volume is chemically prepared to self-ignition, then a detonation wave propagates over the entire space with the sound velocity.

After fire, starting again a shut-down section requires special care: a fire source being latent due to the lack of oxygen may be ignited by the entering fresh air. In order to prevent the destruction effects of a dust explosion, several measures may be done. The first measure is to prevent the build-up of electrostatic charge where dust cloud is present and, especially, where the air motion is slow. For this reason, metal surfaces should securely be bonded to earth. Furthermore, flanged joints of the pipeline should maintain their electric conductivity and the integrity of earthing of all metal parts should be checked from time to time. A well-grounded facility will help to reduce potential dust explosion hazards.

The wood dust and chips are non-conductive materials and they can store some electric charge. Due to frictional charging in the conveying system, particles can carry over the charge into the filter and storage bin generating high electric field. That means that the risk of dust explosion cannot fully be eliminated. Therefore, further measures should be taken to reduce the risk of damages caused by a possible dust explosion.

Filter housing and storage bins, due to their size, are not capable to withstand pressures generated by an explosion. In the case of low risk of explosion, a cost-effective solution may be to use a relief venting. The essence of a relief venting is to accommodate a bursting panel, displacement panel, or a hinged door that operates once and opens a free surface at a given pressure for rapid expansion toward the environment. The size of vent is dependent upon the volume of filter housing or storage bin (EN 1127-1:2011, 8.5).

A more effective solution is the use of detection and suppression system. The starting of an explosion is associated with infrared or visible radiation and pressure rise. It is important to detect the onset of explosion in the early stage, in order to switch off the air supply, to open a venting system automatically, or trigger a suppression system to extinguish the developing flame. At the same time, the plant will rapidly be shut down and no material will be fed into the system.

Dust explosions uniquely could occur in 2 phases, a primary and a secondary, and spread very quickly. The primary one is triggered when dust that has been settling on a surface for a prolonged period of time is suddenly disturbed. The dust becomes airborne in the closed space. When exposed to an ignition source, the dust will be ignited causing the primary explosion. The secondary explosion is often larger and more dangerous. The primary explosion can become an ignition source for the secondary explosion.

In order to use unified approach and well-defined preventing measures, there are national and international standards the use of which are compulsory in the wood industry. In the industry the following practical solution must be used according to the standard EN 12779:2015:

A fire or an explosion within a filter should be detected with suitable sensors (e.g., spark detectors, flame detectors, particle counters, gas sensors for fire, pressure sensors or monitoring of the explosion venting devices, etc.).

Once a sensor detects a fire or an explosion, the following actions will be automatically performed:

- shut down of the chip and dust extraction system;
- disconnection from recirculation air to exhaust air;
- shut down of the filling and discharge system;
- activation of a visual and/or audible alarm.

An explosion within a silo can be detected with suitable sensors (e.g., pressure sensors or monitoring of the explosion venting devices, etc.).

The spreading of fire (heat radiation and flame transmission) from the filter to adjacent equipment and objects should be avoided. Measures against fire spreading are the following.

Ducts connected to the filter (extraction ducts, recirculation air ducts, conveying system) shall be equipped with devices to stop the spreading of fire and smoke.

Extraction ducts and recirculation air ducts should either:

- not pass through a firewall; or
- be equipped with fire gates according to EN 1366-7:2004 (extraction ducts) and fire dampers according to EN 1366-2:2015 or be insulated in accordance with the requirements of EN 1366-1:2014 in the corresponding fire zones.

To limit the effects of an explosion, filters, silos and any other closed storage devices with a storage volume of >1 m^3 shall be equipped with adequate constructive explosion protective devices according to EN 1127-1:2011, 6.5. In cases where extraction pipes need to penetrate fire compartments that have been installed in order to comply with building regulations, an automatic fire barrier that is suitable for pipes carrying materials must be installed. Fire protection flaps in ventilation areas are not compatible with this type of solution.

References

I. Bölöni, A kalapácsos darálók aprítási mechanizmusa (Comminution mechanism of hammer mills). D. of Sc. Dissertation, Budapest (1973)

S. Callé et al., Influence of size distribution and concentration on wood dust explosion. Powder Technol. 144–148 (2005)

EN 149:2002 Respiratory protective devices. Filtering half masks to protect against particles. Requirements, testing, marking (2002)

EN 12779:2015 Safety of woodworking machines. Chip and dust extraction systems with fixed installation. Safety requirements (2015)

EN 1127-1:2011 Explosive atmospheres. Explosion prevention and protection Basic concepts and methodology (2011)

EN 1366-1:2014 Fire resistance tests for service installations, Ventilation ducts (2014)

EN 1366-2:2015 Fire resistance tests for service installations. Fire dampers (2015)

EN 1366-7:2004 Fire resistance tests for service installations. Conveyor systems and their closures (2004)

T. Kauppinen, R. Vincent, T. Liukkonen, M. Grzebyk, A. Kauppinen, I. Welling, P. Arezes, N. Black, F. Bochmann, F. Campelo, M. Costa, G. Elsigan, R. Goeren, A. Kikemenis, H. Kromhout, S. Miguel, D. Mirabelli, R. Mceneany, B. Pesch, N. Plato, V. Schlunssen, J. Schulze, R. Sonntag, V. Verougstraete, M.A. De Vicente, J. Wolf, M. Zimmermann, K. Husgafvelpursiainen, K. Savolainen, Occupational exposure to inhalable wood dust in the member states of the European Union. Ann. Occup. Hyg. **50**(6), 549–561 (2006)

J. Mandryk, K. Udeni Alwis, A.D. Hocking, Work-related symptoms and dose-response relationships for personal exposures and pulmonary function among woodworkers. Am. J. Ind. Med. **35**, 481–490 (1999)

NFPA 654-2020 Standard for the prevention of fire and dust explosions from the manufacturing, processing, and handling of combustible particulate solids (2020 Edition)

NTP Board of Scientific Counselors, FINAL Report on Carcinogens Background document for wood dust. Meeting of the NTP Board of Scientific Counselors Report on Carcinogens Subcommittee, 13–14 Dec (2000), p. 413

Gy. Sitkei, Keverékképzés és égés karburátoros motorokban (Mixture Formation and Combustion in Spark Ignition Engines) (Akadémiai Kiadó, Budapest, 1969)

W&M Respiratory Protection Program (2020), p. 13

Chapter 8
System Optimization

General Remarks

Optimization procedures are important and useful tools in the hand of design engineers for seeking and finding the best possible solutions concerning performance, utility, functional perfection, low energy and operating costs, and economic investment. Before mankind, nature always had a striving after optimum choice in evolution and in the use of its resources. In natural laws, we often recognize proof for it, for example, the principle of minimum potential energy. The mankind had similar striving from ancient time.

The history recorded two famous "iso-perimeter problems": Queen Dido of Carthago received a land that she could fence around with the skin of an ox cut into fein ribbons (around 825 B.C.), and the Roman hero, Horatius Cocles, received as much land as he could plow around in a day (around 500 B.C.). Zenodoros (200 B.C.) had already proved in logical way that the optimum choice, concerning maximum land area, is the circle or semi-circle. Much later, in the eighteenth century, J. Bernoulli, Euler, and Lagrange paved the road to develop exact mathematical methods (Szabó 1987).

Any optimization method consists of a performance measure (objective function or functions) with independent variables, a set of equality, and inequality constraints that constitutes the model of the system. In formulating the model, it is necessary to select functional relationships describing the system behavior as a function of independent variables. The objective functions and the constraints are generally nonlinear. The objective functions are related to performance and economy. Such performance characteristics are the production rate, production time, quality assurance, and product utility value. The problems of economy cover the minimum production cost, minimum energy and raw material consumption, optimum use of machine capacity, utilize the space within a building, minimize investment costs, maintenance cost, minimum pollution of workroom and environment, etc.

E. Magoss et al., *Dust Extraction and Handling in the Wood Industry*,
SpringerBriefs in Applied Sciences and Technology,
https://doi.org/10.1007/978-3-031-08915-2_8

The objective function is formulated in accordance with the selected criterion on the basis of which the performance or economy can be evaluated. The different criteria may be in conflict with each other. It is not possible to simultaneously maximize or minimize several criteria. If we make multi-criteria optimization with several objective functions (Pareto optimum) then we need to make a compromise in selection of a final common optimum. In practice, it is purposeful to select one main criterion and the other criteria are treated as constraints having prescribed maximum or minimum values.

8.1 Methods of Optimization

The importance of optimization in all fields of technical activity was always recognized, at the same time, the successful applications are less frequent. Many reasons may be mentioned which hinder the everyday use of this useful method.

A successful optimization procedure presumes many fundamentals, physical knowledge, and manageable mathematical method for the practicing engineers. Especially important are the following:

- knowledge of the given physical process or machine design requirements,
- availability of functional relationships describing system behavior as a function of influencing variables,
- a quantitative criterion, an objective function should be selected,
- the necessary constraints should carefully be selected,
- if it is possible, the fundamental constraint is to be determined,
- examine the feasible range of influencing variables. Variables with narrow feasible range will not influence the optimum,
- choose the appropriate mathematical method.

There are two reliable mathematical methods for optimization:

- strict non-linear mathematical method using, for example, the Fletcher-Powell algorithm for solving the objective function with a set of equality and inequality constraints (Csanády et al. 2019),
- a newly developed and used engineering optimization method based on the recognition of the main constraint. This method, in many cases, allows to derive generally valid analytical solutions (Csanády et al. 2019).

In both cases, the availability of functional relationships is indispensable.

8.2 Characteristic Features of Optimization in Dust Extraction and Handling

The dust extraction and handling process has many unusual features which fundamentally influence the use of any optimization procedure. These characteristic features are the followings:

- the minimum air flow rate (m^3/h) for machine hoods is prescribed by the manufacturers and they are fixed input data, improvements on suction hood efficiency is generally falling out of the users' activity,
- the minimum air transport velocity is well established as a function of transport concentration and dust properties. A variation is possible only upwards which is unfavorable in all aspects,
- the allowable dust concentration in workroom is determined by national and international standards and it is also a fix datum,
- the transport concentration is determined by the operating parameters of woodworking machines which is independent of the extraction system,
- the energy consumption of extraction and conveying system is relatively high but it can hardly be modified by varying system variables.

The above-mentioned features fundamentally hinder to carry out a classical optimization procedure. In this case, taken the allowable minimum air velocities and the prescribed air flow rate, a direct calculation method supplies the same result in a much shorter time.

In the design phase of a dust extraction and conveying system one may encounter more problems with the optimal location of machines concerning internal material transport and the dust extraction conveying pipelines. Latter should be as short as a possible reducing investment costs and also energy consumption in a certain extent. A further problem is the varying operational condition when the number of machines is not constant (see Sect. 6.3).

Dust getting out from the machine hoods fundamentally contributes to the workroom pollution. In order to ensure the allowable dust concentration, it would be not favorable to use less air volume than prescribed by the manufacturer, because this would increase the required air volume of ventilation. Further developments in suction hood constructions may reduce the direct pollution of workroom and herewith the required ventilation air.

The use of a pre-separation device needs additional investment cost but it may pay itself reducing the ventilating air considerably. The use of the collected dust may influence the required volume of storage equipment and a longer time of storage makes it necessary to use flow aid devices for unloading. A flow aid device means additional cost but it reduces the height of the silo which may compensate the total cost compared to a silo with hopper outflow.

It may be concluded that more possibilities are given in the design phase to optimize the whole production system. The dust extraction and conveying system is an indispensable serving facility with a required high degree of reliability.

References

E. Csanády et al., *Optimum Design and Manufacture of Wood Products* (Springer, 2019)
I. Szabó, *Geschichte der mechanischen Prinzipien* (Birkhäuser Verlag, 1987)

Index

E. Magoss et al., *Dust Extraction and Handling in the Wood Industry*,
SpringerBriefs in Applied Sciences and Technology,
https://doi.org/10.1007/978-3-031-08915-2

The manufacturer's authorised representative in the EU is Springer Nature Customer Service Centre GmbH, Europaplatz 3, 69115 Heidelberg, Germany. If you have any concerns regarding our products, please contact ProductSafety@springernature.com

Printed and bound by CPI Group (UK) Ltd, Croydon, CR0 4YY
28/09/2026
02224079-0005